EGG
&
EGO

Springer
New York
Berlin
Heidelberg
Barcelona
Hong Kong
London
Milan
Paris
Singapore
Tokyo

EGG & EGO

An Almost True Story
of Life in the Biology Lab

J. M. W. SLACK

Springer

J.M.W. Slack
Department of Biology and Biochemistry
University of Bath
Bath BA2 7AY
UK

Library of Congress Cataloging-in-Publication Data

Slack, J.M.W. (Jonathan Michael Wyndham), 1949–
    Egg and ego : an almost true story of life in the biology lab /
J.M.W. Slack.
        p.   cm.
    Includes bibliographical references and index.
    ISBN 0-387-98560-3 (pbk. : alk. paper).—ISBN 0-387-98559-X
(hardcover : alk. paper)
        1. Science—Humor.   2. Biology—Research—Humor.   I. Title.
PN6231.S4S57   1998
818'.5407—dc21   98-7732

Printed on acid-free paper.

Production coordinated by Impressions Book and Journal Services, Inc., and managed by
    Steven Pisano; manufacturing supervised by Jeffrey Taub.
Typeset by Impressions Book and Journal Services, Inc., Madison, WI.
Printed and bound by R.R. Donnelley and Sons, Harrisonburg, VA.
Printed in the United States of America.

9 8 7 6 5 4 3 2 1

ISBN 0-387-98560-3 Springer-Verlag New York Berlin Heidelberg   SPIN 10680941   (softcover)
ISBN 0-387-98559-X Springer-Verlag New York Berlin Heidelberg   SPIN 10680967   (hardcover)

To Janet, Becky and Pippa

Contents

Prologue: Cosmic Egg and Human Ego

This is a book about scientists and what it is like to be a scientist today. More specifically, it is about life in the biology lab in the era of genetic engineering. The public is often fearful of the closed world of the genetic engineers, so I have tried to lift the curtain that shrouds it—to go inside the laboratory and expose the hopes, fears, and motivations of the people who live there. When we look at what the scientists are doing, we find it is not so strange after all. Their world is a competitive struggle to get money for research, another struggle to get their experiments to work, and yet another struggle to publish the results in the most fashionable journals. In these competitions, many fall by the wayside, but a few succeed and become stars, adored by their specialist public much as successful actors, musicians, or sportspersons are admired by wider society. Indeed, the egos of the research stars are just as large as those of the better-known professions, and they behave in ways quite recognisable to outsiders, with a familiar personal motivation toward publicity and career development.

At the same time, I have tried to explain some recent advances in biology and the excitement they have generated. The "egg" in my title is partly a real egg, the subject of our research, and partly a cosmic egg, representing the great potency of science. Science is full of very specialised studies on very small, detailed questions. Just occasionally, the answer to one of these questions behaves like an egg. It hatches into a wholly new field of knowledge that grows exponentially and generates enormous changes for academic science and, a little later, enormous practical benefits—and occasionally problems—for the wider society.

The book starts in Chapter 1 with a particular experiment I conducted in 1986. I am a developmental biologist—that is, one who wishes to understand how living organisms develop from eggs or seeds—and the experiment concerned the identification of a key substance controlling animal development. The result of the experiment was an important one in terms of shaping my own career and work. In other chapters (3, 5, and 8) I explore some of the ramifications of three scientific fields that my own work has impinged on: the biochemistry of growth factors, the developmental biology of early embryos, and the evolution of developmental mechanisms in the history of life. These chapters explain some of the key ideas and conclusions reached in recent years and try to communicate some of the excitement felt by those engaged in the work.

These chapters represent the egg. But science is produced not just through the procedures of hypothesis and experiment, important though these are. Science has its own living, working environment with its own culture; an examination of this brings us to the ego. To understand what it is like to be a scientist, it is just as important to appreciate people's motivations and behaviour as the formal aspects of their work. There are many books on the philosophy of science, but perhaps because these are mostly written by nonscientists, they do not give any idea of the actual day-to-day nature of the work. I feel it is important to remedy this deficiency, for the benefit both of students contemplating a career in the life sciences and of those in the general public interested in what science is really like. So the remaining chapters of the book deal with various incidents I have seen, heard of, or been a part of myself. In these sections I have perhaps emphasised the dramatic and played down the long periods of relatively humdrum activity that exist for any profession, but it is nonetheless a story of real life in the biological laboratory. It describes the system of academic science—where the money comes from, the career paths that people follow, and the enduring vanities of scientists that make them resemble actors as much as any other profession. My experience is mostly in the United Kingdom, but science is international in character, and the differences between the United Kingdom and the United States are much less significant than are the similarities.

All the events recounted really happened, or are at least thought by someone to have happened. I have, however, transposed a few events in time and space in order to make the story flow more smoothly. I have also in these chapters altered the names of certain individuals and institutions to protect them from embarrassment. In this category are James Samson, Tobias Fortune, Jack Large, John Field, George, Ben, Sylvia, Jaspar, and the professors at the University of Portree. Indeed the "Serious Disease Society," and the "University of Portree," do not really exist, although they have a certain resemblance to real institutions. In a few cases where individuals might be seriously embarrassed, they have been "composited"; that is, they are not single people but are instead mixed characters possessing attributes drawn from several different real people. This is true of Messrs. Simon Law, Bob Franklin, and Mehmet Hussein. It is in these respects that my narrative is an *almost true* rather than a *literally true* story of life in the biological laboratory.

My message is that science is indeed a fusion of the egg and the ego, the boundless promise of the small new idea combined with the vanities and competition of ordinary human beings. It can only be fully understood if this dual character is appreciated. Lest some be shocked by what they read, I want to emphasise that I have no intention of putting off able people who want to enter science. All professions have their trials and tribulations, and this is no exception. Academic science does share with acting a high drop-out rate, as there are many more entrants than the system can possibly accommodate in the long run. However, most of those who leave go to do something else that is perfectly valuable and worthwhile; and there is everything to be said for ensuring that at least some people in other areas of life, such as politics, business, or teaching, know what science is like on the inside. At the same time, academic science will never be short of recruits. There is a magic to the well-executed experiment, and the occasional experience of it is enough to drive people on and sustain them through all the failures and disappointments that consume most of their time.

I also do not wish to undermine the validity of science. It is true that the dry, formal, deductive character of science is a fiction. It is maintained largely because scientists and nonscientists alike seem to wish to maintain that science has some special status that guarantees the truth of its findings. Any hint that values might enter science from outside sources, such as economic imperatives or the fashions of wider society, are strongly resisted because it is felt that the recognition of such influences would weaken the authority of the results. It is also true that the modern practitioners of "science studies" will use such examples to argue that the profession of science is merely a priesthood and that there is no more value in molecular biology than in astrology or homeopathy. This is not the place to pursue this argument in detail, but lest my book be misunderstood, I want to put on record that I do believe that the results of science are on the whole correct and valid. I think it is difficult to identify anything that uniquely distinguishes scientific knowledge from other forms of knowledge; the ultimate justification of value should be sought in successful practical application rather than in philosophical debate. The practical applications of the fields I shall discuss in this book are just beginning to be realised. But when they are fully realised, they will change the world profoundly.

J.M.W. Slack

The Experiment

They looked like tiny sausages, and never had I been so pleased to see a sausage. They were actually little pieces cut out from frog embryos, so small that they could only be seen clearly down a microscope. Normally, such small clumps of cells will round up into balls and stay that way for several days, remaining alive but in a very inscrutable way, as they do not do anything much to attract the observer's attention. But the ones I was looking at had elongated into sausages. I knew that their unusual shape was due to their having been treated with a protein we had purified from cows' brains. I also knew that the elongation I saw meant that the cells were developing in a different way from usual and that they would decisively change the course of my future work.

This was an experiment. The treated tissue explants were the experimental cases, and the untreated ones, taken from the same embryos on the same day and kept in the same salt solution, were the controls. Described like this, the experiment may not sound either interesting or impressive. But it actually marked the convergence of two lines of scientific work both extending back about 50 years. One was the science of experimental embryology, which asks: "How do embryos develop?" The other was the study of tissue culture, or the growth of cells in tubes or bottles outside the body. Our experiment had shown that one of the mysterious *inducing factors,* the substances that control the development of the embryo, was the same as one of the *growth factors,* the substances that control the multiplication of cells in tissue culture. The excitement arose because we knew that people would generalise from this and say: "Inducing factors are growth factors." Because a lot was already known about growth factors, it meant that we now knew a lot about inducing factors as well. In fact, we knew that neither experimental embryology nor growth factor biology would ever be the same again.

Brains

Before my discourse rambles too far into the realms of hard science, careers, personal fame, and money, I should put some flesh on what we actually did. Flesh is the appropriate word here, as our experiment started in a slaughterhouse. We all know that slaughterhouses exist, but few of us ever visit one. In fact, apart from farmers, butchers, and meat inspectors, one of the few groups of people who ever visit slaughterhouses are biochemists. They go to get fresh animal tissues as a starting point for the purification of their favourite enzyme or hormone. Although I have a degree in biochemistry, I am not really a biochemist, and the trips we made to get the brains are my only personal experience of this type of expedition. The year was 1986, and what we wanted was a substance called *fibroblast growth factor* (FGF). FGF had recently been purified and characterized in America, and we wanted large amounts for our embryo experiments. The reader must bear in mind that everything is relative and that in biochemistry a "large amount" of something means about a milligram, a quantity weighing as much as a cube of water one millimetre on each side. But because growth factors are fabulously potent substances, they occur in animal tissues only in very small quantities. Compared with most other animal tissues, brain is a good source for the extraction of FGF. Even so, to obtain a milligram of FGF, we knew we should need at least a kilo of tissue; to be on the safe side, we decided to collect 10 kilos.

At the time I was working in Oxford, England, and it turned out that the only slaughterhouse near the university was located on the industrial estate of the village I myself lived in. So I stayed at home that morning and kept phoning to find out when the cows would be going through. This brought the first problem. Slaughterhouse men do not work to much of a timetable. They just kill and process whatever drives up, so they have no idea whether cows, pigs, or sheep would be coming off the line in a couple of hours time. There was also a slight communication problem because a city-reared individual like myself has only a dim idea about the subdivisions of domestic animals. So if I asked whether they were likely to be doing any cows in the next couple of hours, they would say, "Probably not, because we've got the heifers to do first." After a few phone calls dogged by mutual incomprehension, I summoned up my resolve and called out the remainder of the party from Oxford with their wellies, rubber gloves, and buckets of dry ice. We entered the slaughterhouse with some apprehension, steeling ourselves to the sight of the carcasses whirling around overhead on the conveyor belts. You have to move around quite carefully in this sort of environment; otherwise you can be struck a mighty blow from behind by an upside-down pig.

Fortunately for us, they had in fact just finished doing some cows (or maybe it was heifers), and we gingerly asked if we could have some heads please. We agreed the price—one pound sterling per brain, in cash—and we were gestured into a side room where there was a heap of fresh heads and a bandsaw. My colleague John Heath, who is a past-master at this sort of thing, seized a head and showed us how to cut it in two down the midline (i.e., separating the right and left sides). "There's the pituitary," he said to me, pointing at a pea-sized object at the base of the brain. Pituitary glands have a special place in biochemical culture because they are the source of a wide variety of important hormones and other bioactive molecules. Some substances require tons of pituitaries for their purification, and there are companies that scour the whole United States collecting vast quantities of the tiny organs for the use of biochemists. We could have used pituitaries, which are an even better source of FGF than brain, and perhaps 500 grams would have been enough for our requirements. But we knew that we could not collect the quantities required, since 500 grams of pituitaries would mean 500 cows.

"Oi! What are you doing?" said one of the slaughterhouse men, bringing the anatomical demonstration to an abrupt close. In the ensuing discussion, it turned out that the heads are not entirely a waste product of meat processing. They become the property of the slaughterhouse men themselves, who sell them on to make glue or fertilizer. Although it didn't seem to us to matter much whether one whole or two half heads was made into glue, the men obviously felt that the acceptability of their product depended on its appearance. In short, the anatomically correct midline section into right and left halves was not at all correct from the financial point of view. However, this presented us with the obvious problem of how we could acquire our brains while leaving the heads entirely intact. The ancient Egyptians are said to have removed the brains from their corpses before mummification and to have done so using small hooks pushed through the nose. Modern otorhinolaryngologists can also remove your pituitary by the same route if it should become necessary. But to take such care over a dismembered cow's head was surely going too far. We agreed to compromise by sawing a small piece off the top of each skull and removing the brain by hand, thus leaving the heads looking more or less normal but minus their brains. Each brain weighs about 400 grams, so we decided to do 25 heads to get our 10 kilos. We wrapped them in foil and froze them in dry ice.

John Heath was an essential element in the design of the experiment. He is tall and handsome in a mustachioed sort of way and radiates confidence about obtaining money and equipment and the other tiresome necessities of laboratory life. He lives by the broad sweep of generalisation and had early on understood that growth factors were going to be really important. His upbeat style and willingness to assume the mantle of "Mr. Growth Factor" had acquired him a range of lucrative consultancies with the pharmaceutical industry. As we set off on the brain-collecting expedition, he remarked that the nearest slaughterhouse to Oxford would certainly be an academic sort of place. Doubtless, as one peered through the rows of carcasses swinging on the conveyor, one would glimpse Nobel Prize winners at their work. Undoubtedly, there would be a seminar room with slide projector and coffee-making facilities to enable these august visitors to communicate their latest results while they were waiting for the cows (or was it heifers?). He was wrong. It was just an ordinary slaughterhouse, one of the many in Britain struggling to meet the current European Community standards of hygiene and cleanliness. A couple of years later, it had closed down, little realising the importance it had held in the biochemical culture of Oxford University.

By the time we had collected enough brains, it was about lunchtime. So we went up to my house for a sandwich, leaving the containers of brains in the garden shed for safety. "Where are the brains, Dad?" asked my three-year-old daughter, who knew something unusual was going on. We ate our cheese sandwiches, loaded the brains into our cars, and set off back to Oxford.

In the Cold

There is in every biological laboratory a place called the "cold room." This is a whole room that is cooled to about 4 degrees centigrade, the temperature of a domestic refrigerator. Once upon a time, such places were regularly used for the purification of enzymes from animal tissues, the idea being that the low temperature of everything slowed down the deterioration of the tissue and preserved for longer the enzyme of interest. But the great days of biochemistry are long gone. Nowadays enzymes are mainly purchased from chemical companies, and cold rooms are used mainly for storage. Being communal areas, they inevitably accumulate large amounts of rubbish: forgotten agar plates, bottles of media, racks of test tubes. The existence of such places for many years seems also to have allowed the evolution of a pervasive type of black mould that infests cold rooms and rapidly coats the surfaces of any objects that are left there.

We went to work with the disinfectant and black refuse bags and cleared all the rubbish away from a sufficient area to process our brains. The first part of the preparation involved extracting the water-soluble proteins by homogenizing the brains in a device similar to a domestic food mixer and stirring the resulting soup for a while in huge buckets with a powerful paddle stirrer. The soup was then put in tubes and centrifuged in a large refrigerated centrifuge to remove all the insoluble material, which forms a pellet at the bottom of the tube, from the soluble components, which remain in the bulk liquid, otherwise known as the *supernatant*. This supernatant is a highly complex mixture containing many thousands of substances, including the FGF that we wanted. The first purification step was the addition of many kilos of ammonium sulphate, which precipitates different groups of proteins at different concentrations, followed by further rounds of centrifugation to remove the precipitate thrown down after the addition of each batch of ammonium sulphate. The main problem for us at this stage was physically handling the 10 kilos of brain, which rapidly became converted into about 100 litres of slimy, smelly homogenate. All the tubes and bottles in the lab were far too small for such quantities, so we were constantly running backwards and forwards to the blender or the centrifuge and constantly attempting to find some acceptable way of disposing of the huge amounts of brain sludge that accumulated at the bottoms of the centrifuge tubes.

Fortunately at this time (1986), we had yet to hear of bovine spongiform encephalopathy (BSE), the "Mad Cow Disease" that was to sweep across Britain in the following years. This disease, which later became widespread in British cattle, has a long incubation period, but once it develops, it leads to rapid, irreversible, and fatal degeneration of the brain. Farmers in the 1980s regarded the disease as a minor irritation. In the early 1990s, BSE developed into a massive epidemic. The Ministry of Agriculture loudly declared that "British beef is safe." Foreign countries, whose wallets were safe but who valued the brains of their own citizens, banned the import of British beef. In 1996 a new form of the human Creutzfeldt-Jakob disease (CJD) was described, which was unique to Britain and very similar in all respects to BSE. Both BSE and CJD are now known to be caused by a particular protein found normally in the brain, called a *prion*.

This protein can change its shape to an abnormal form that can then catalyse the conversion of more of the normal to the abnormal form. The abnormal form builds up to a high level and kills the cells in which it is accumulating, producing "spongiform lesions" and slowly killing the patient. Although it is still unclear whether the disease can really be transmitted by eating beef, a surprising number of biologists, veterinarians, and others in the know seem to have stopped eating beef at some time in the intervening 10 years. We certainly would no longer be permitted to wallow around in brain homogenate for several hours as we did during the FGF preparation. Whether one can catch BSE by breathing in brain aerosol we do not know. But we are all still here, and the degeneration of our brains is no worse than might be expected 13 years on, so maybe our cows were free from infection, or maybe we were lucky.

After the blending, centrifugation, and precipitation with ammonium sulphate, the preparation was redissolved in water and put into celluloid bags resembling giant condoms, then hung in a large stirred container of water. This step, known as *dialysis*, removes the excess ammonium sulphate, which diffuses through the celluloid bag, leaving the proteins inside. Because of the large amount of ammonium sulphate inside the bags at the start, they tend to absorb water and swell. Around midnight, we nervously inspected the tense-looking tubes: "Will they burst? Is it safe to go home?" Success in science goes only to those who are utterly paranoid about things going wrong and utterly obsessive about preventing mishaps. We untied all the bags, distributed their contents into more bags, and hung them back again in the 40-litre tub of stirred water. About one o'clock in the morning we staggered off to bed. When I got up the next morning, my little daughter asked me: "Dad, where are the brains?"

The bags didn't burst, and we were able to resume the preparation the next day. By this stage, it had become relatively civilized, as we were dealing with just a cupful of relatively clear and not too foul-smelling fluid. This now needed to be processed through two steps of column chromatography. Columns are the essence of protein biochemistry, and all biochemists' spouses must be sick of hearing about columns not working, running dry, clogging up, and so forth. Spouses are also aware that columns keep curious hours and often seem to need attention in the middle of the night. A column actually consists of a glass, plastic, or metal tube containing some material that absorbs different proteins under different conditions. The mixture to be purified is first pumped on to the column. Some components stick to it and others pass straight through. Then a different solvent is pumped through, which makes some of the proteins that were sticking detach from the column and come out in the flow of solvent. The flow leaving the column is collected in numerous test tubes, and the contents of each tube is called a *fraction*. A good column will concentrate all of the protein you want into a small group of fractions, accompanied by the minimum quantity of other proteins. There are numerous types of columns, and numerous combinations of solvents for loading and eluting columns. The fancier systems are machines that operate rapidly and at high pressure, and these are very expensive, easily costing $50,000.

Our preparation required a column containing a substance called heparin, best known for its ability to prevent the clotting of blood, but also able to bind FGF very strongly. This was followed by an ion-exchange column that separates proteins according to the electric charges on the molecules. After this, we had to run a sample on a "gel" to establish its purity. Gels are even more ubiquitous than columns in the modern biochemical laboratory, as they are the main technique for separating molecules on a small scale. They are literally slabs of jelly across which an electric field is passed. The electric field makes each type of substance migrate across the gel at a different rate, so a complex mixture resolves into a row of bands, with each band consisting of a single substance. Our gel showed just one band, so we knew that our FGF preparation was pure. Purity is very important, because if your substance is pure you know that any biological effects it has are really due to the substance itself and not to some minor contaminant or impurity. In fact, the whole object of our experiment was to be able to look at the inducing activity of a guaranteed pure substance rather than a crude tissue extract.

About two weeks after the visit to the slaughterhouse, we had our little tube of FGF ready for use. Now the real panic started. Would it actually work? Had we spent all this time for nothing? We had to test our FGF by putting it onto the tiny pieces of tissue taken from frog embryos. What we were looking for was a change in the developmental fate of these tissue fragments. Without treatment they would form skin cells, but we were looking for something else, particularly the formation of muscle.

The Result

When an embryo develops, the fertilized egg divides many times to form a ball of cells called a *blastula,* and these cells differentiate from each other to form the different tissues of the adult body (skin, nerve, muscle, and bone). In the early years of the twentieth century, experimental embryologists had found that the type of differentiation followed by a group of cells often depended on the nature of the neighbouring tissues. In such cases, the neighbouring tissues were presumed to be emitting chemical signals that could control the pathway of differentiation, and these hypothetical substances were called *inducing factors.* The embryos of the newt or the frog have many advantages for experimental work and have long been used by embryologists to tackle a whole variety of problems. One reason for this is that it is easy to graft things, either living tissues or test substances, into the embryos. The surroundings of a graft are influenced by any inducing factors emitted by the graft, and a great deal of work had been done in attempts to characterize some of the factors, although none had yet been identified as a particular single substance.

Nearly all types of animal, including humans, have an embryo that consists at an early stage of three tissue layers, called *ectoderm, mesoderm,* and *endoderm.* Roughly speaking, the ectoderm forms the skin and nervous system, the mesoderm forms the muscles and skeleton, and the endoderm forms the organs of the gut and respiratory system. Because these three tissue layers are found in all animals, the mechanism of formation for any one of them is an issue of considerable scientific importance. Back in 1939, the first paper had appeared describing the induction of mesoderm by small pieces of guinea pig bone marrow that had been implanted into early amphibian embryos. So it was known that there was a type of substance, present in certain adult animal tissues, that could cause mesoderm to arise from regions of the early embryo that did not normally form mesoderm. Some unfortunates in Germany had attempted for decades to purify these substances, but without success, as they are very active and present in very small amounts. The technology of protein chemistry (i.e., columns with associated pumps and detectors) did not become equal to their purification until the mid-1980s.

Also in 1939 a paper appeared from another laboratory, in another country, describing how an extract of brain could support the growth of fibroblasts in tissue culture. Fibroblasts are a kind of cell derived from the connective tissue of the body, and as their name suggests, they secrete "fibres" of the insoluble protein collagen. They are quite easy to grow in culture, but like almost all tissue culture cells, they will not do so unless the liquid they are growing in contains some complex component such as animal serum or embryo extract or brain extract. During the 1950s and 1960s, it had gradually became clear that these extracts were the source of several growth promoting factors, present at fabulously low concentrations, but with fantastic biological activity. Slowly but surely the growth factors were purified and characterized. The growth factor that had first been discovered in brain extract in 1939 became known as fibroblast growth factor (FGF) and was first purified in the mid-1980s.

By now readers will see clearly the nature of our experiment. We had prepared some pure FGF. We had put it onto pieces of frog embryo from a region that did not normally form mesoderm, and the change in shape of the tiny tissue explants, from balls to sausages, told us that mesoderm had been induced. So FGF was a mesoderm inducing factor. More generally, it now seemed likely that growth factors and inducing factors were not different types of substances doing different jobs but were actually one and the same thing.

A few years before, there had been great excitement in another field of biology, that of cancer research, due to the discovery of *oncogenes*. Oncogenes were bits of DNA—literally, isolated genes—that, if introduced into otherwise normal cells growing in tissue culture, could make them behave like cancer cells. There had been a great race to identify the oncogenes, and the very first one to be identified had been shown, three years before our experiment, to code for a growth factor. Because of the work in tissue culture, we already knew that growth factors could make cells grow in bottles. Because of the oncogene story, we knew that they could cause cancer. Now we knew that they could also control the development of the embryo.

So What Is an Experiment?

In principle, an experiment is a question put to nature. It involves a particular hypothesis: that some entity exists, or something affects something else in a predicted way. Much ingenuity may be expended turning such a general hypothesis into a specific prediction that can be tested by an experiment; in fact, the intellectual side of science consists mainly of designing such experiments. A successful experiment will test the prediction and lead to a clear confirmation or rejection of the underlying hypothesis. Although scientists do not like seeing their pet theories disproved, they will usually accept the results eventually, and a willingness to be proved wrong and to abandon cherished ideas must be counted as one of the greatest strengths of science.

An experiment should be set up in such a way as to isolate just the particular variable under consideration from all the other thousands of possible influences. In practice most experiments are extremely mundane, and the aspect of nature under investigation lies within the technology of the lab itself. For example, many procedures in molecular biology are long and complex and involve the use of several enzymes that cut, join, or otherwise modify pieces of DNA. If some procedure of this sort has not worked, as happens all too commonly, the investigator may frame the hypothesis: "My batch of enzyme has gone off." He will test this hypothesis by setting up two identical test tubes containing all the things his enzyme needs to show its activity. To one tube he will add the suspect enzyme, to the other the same amount of the same enzyme but taken from a different batch or different manufacturer. The first tube is the called the experimental tube and the other is called the control tube. If the enzyme in the control tube catalyses the reaction as expected, this shows that all the other components of the reaction mixture are satisfactory. If the reaction in the experimental tube does not work, the investigator will conclude that his enzyme had indeed gone off. He will throw it away, replace it with a working batch, and move on to the next experiment.

This simple example is fairly typical of biological experiments because the design involves a comparison between a test procedure and a control procedure, predicted to differ only in respect of the variable under study. In fact, there are often two controls: a positive control predicted to show the effect and a negative control predicted not to show it. In our example the control was a positive one, namely, the enzyme known to work. The investigator may also have included a negative control, which would be a third tube containing the same reaction mixture but without the enzyme at all. In this tube we should expect the reaction not to proceed, which would confirm that the presence of the enzyme was in fact necessary for the reaction to occur. The example is also typical in that the result is of absolutely no interest to anyone but the investigator himself. But every so often, we are able to manoeuvre ourselves into a position where we can really put a question to nature, and this sort of experiment is more exciting.

In our real experiment, the hypothesis was that FGF could induce mesoderm. The specific prediction was that the treated embryo explants would elongate and form tissues such as muscle. The control was a negative one: the untreated tissue explants, which remained spherical and which developed into skin cells as usual. We also often included a positive control, which was a preparation already known to induce mesoderm. At the time this would have been some mixture of unknown composition, such as a chick embryo extract or cell culture supernatant. Of course, the visible elongation of the tissue explants alone was not sufficient evidence of mesoderm induction. It was also necessary to make further observations to confirm the presence of mesoderm-derived tissues in the explants. This was done in two ways. First, we would use a machine called a microtome to cut the explants into very thin slices suitable for staining and examination down the microscope. Second, we would analyse some of them for the presence of proteins found only in muscle, such as myosin.

A successful experiment on a significant problem gives one the privileged feeling of having lifted a small corner of the curtain that covers the workings of nature. Many scientists never experience that sense of awe. A few are lucky enough to experience it many times. I have felt it only the once. But it is a strong drug, and even one short taste is enough to make one an addict for life. The awe is due to the feeling that things can never be the same again because of what you have seen. But in order for your druglike perception to change the world, much more is required than the fact that you have seen and felt it. It requires that you be listened to and taken seriously, that the technical base of other institutions is adequate to repeat and extend the work, and, above all, that someone else is prepared to pay for it all. In other words, even the most "pure" and "basic" of scientific research is carried out in an institutional framework that has its own laws and its own sociology. Our experiment will serve as an introduction to this world.

The Greasy Pole

Modern biological science has some resemblance to life on the western front during World War I. It keeps going by pouring vast masses of cheap cannon fodder into the no-man's-land of frontline research. Many fall, but some survive and become Ph.D. graduates. The Ph.D., or doctor of philosophy degree, is the entry qualification for professional scientists and is something you do after you have finished your B.Sc. In the United States it takes at least five years to get a Ph.D., and in continental Europe probably no less. In Britain we specialise in a sort of cut price Ph.D. that takes just three years. But those three years are spent entirely in the lab, working on a research project. There is no time for any more formal course work, so British scientists tend to be rather less well informed than their American counterparts. Despite this, Britain seems to be able to go on producing at least some good scientists. So perhaps being well informed doesn't matter as much as we like to think.

Unlike the qualifications of accountants, solicitors, or medical doctors, a Ph.D. does not by any means guarantee a job, and most Ph.D. graduates will not ultimately stay in scientific research because there are far too many of them to fill the jobs available. Getting a permanent job in academic science depends partly on ability, partly on jumping through a number of hoops, but mainly on luck. A Ph.D. does have the distinction of entitling you to call yourself "doctor." But the qualification is still relatively unfamiliar to the general public, and they usually assume that "Dr." means medical doctor. The resulting embarrassment means that few Ph.D.s actually call themselves "Dr." unless they are applying for a mortgage to buy a house, or putting themselves up for election to Parliament.

Edinburgh

I had studied biochemistry for my first degree and decided that I wanted to do a Ph.D. in something involving nucleic acids, because it was clear even then that they were the molecules of the future. I did it at Edinburgh University in Scotland. To go there had been an impulsive decision based mainly on the sight, from the Edinburgh Zoology Department, of some impressive cliffs called the Salisbury Crags in one direction and the sweeping scenery of the Braid Hills in the other. In those days, being young and revolutionary, I was unimpeded by considerations of career success and did not take much advice about where to go. Nowadays you are assumed to have sunk without trace unless you go to a "good place" to do your Ph.D. Good places are identifiable mainly because they are populated by "good people," and good people are identifiable because they publish their results in "good journals." The importance of the good journal is explained in Chapter 4, and there is a further exploration of the concept of the good place in Chapter 9. But for now, the main point to notice is that successful scientists command a hero worship from their specialist public in the same sort of way as do successful actors or musicians from the general public. When a future interviewer looks at your curriculum vitae (CV), he will say with admiration, "So you worked with Zorro the Great," and in an aside to his colleagues, "He must be all right; after all he has been near Zorro for three years." On the other hand, if you have done your Ph.D. in the laboratory of some nonentity, you have some catching up to do with regard to credibility, and in the present rat race, you may never catch up.

My supervisor had been moderately well known in his day, but the time that I arrived in Edinburgh coincided with the time that he seemed to lose interest in science. Whether this was a consequence of my arrival or entirely coincidental, I cannot say. He had just purchased a large, crumbling stone barn in the countryside and had set out to rebuild and refurbish it entirely with his own hands. This naturally meant that he was not around in the lab very much, although he did consent to discuss my results at his home every few months. He was a quiet and thoughtful man and a meticulously tidy worker. Everything would be ready and laid out before he started a procedure, there would be no mess, and all his movements were careful and deliberate. He imparted these virtues to me in the course of some early demonstrations before his house-building activities became completely full-time. He really should have been an inventor as he was very good at designing new pieces of equipment and would sometimes actually make them himself. Unfortunately for him, this was the early 1970s, and British universities had yet to be led to the marketplace. Nowadays, after the long Thatcher terror, anything remotely useful that a university faculty member might think of is immediately seized upon and patented by the university's technology transfer organisation. But most of his ideas languished in his notebook, or as half-built prototypes. He was also unfortunate because in the world of molecular biology, nobody makes anything for themselves anymore. The types of equipment we use are very standardised and are all available from suppliers at lower price and better quality than a departmental workshop can make.

I had chosen my lab with the naive idea that, because it was working on several different sorts of animal and plant, I would have a wide choice of projects to work on. In reality, as I learned later, postgraduate students should not be allowed to choose their own projects or otherwise flounder around on their own. They have to be given a project and made to do it. But as it turned out, I was one of the last beneficiaries (or victims) of the old "sink-or-swim" school.

Sink-or-swim used to be common practice in Britain as it fitted in well with the old idea that the main function of a college tutor at Oxford or Cambridge is to identify the "first-rate chap," and I use the word *chap* rather than *person* deliberately. The tutor would sit and observe the students passing through his care until he spotted a first-rate chap. This chap would be specially groomed and duly installed in a college junior fellowship. He would eventually become a tutorial fellow himself and spend the remainder of his career looking out for another first-rate chap who would succeed him in his turn. This system maintained a self-appointed academic élite in Oxford and Cambridge, and because first-rate chaps are few and far between, it also maintained a steady state career structure in which each cohort of first raters succeeded the previous one.

As applied to the lab, this system meant that the student was left entirely alone. If he, or occasionally she, had talent, he would produce a worthwhile piece of work, and if not he would drop out. There is no point, according to this theory, in rescuing the dropout because he would not have been a first-rate scientist anyway, and if he is only going to be second rate, then who cares? The sink-or-swim philosophy could still flourish in the balmy days of the 1950s and 1960s when there was plenty of state funding for education and science. It is harder to justify in the hungry 1980s and 1990s, when every penny of public money must be accounted for and when success in research has to be guaranteed in advance.

Most postgraduate students in Britain are supported financially by the Research Councils, the state funded bodies whose function is to look after basic research. The pay is very low, and students themselves accord higher status to research assistantships, which are salaried posts attached to research grants, from which it is often possible to register for a Ph.D. Interestingly enough, the lab heads hold research assistantships in lower regard than studentships because they are seen as "jobs" rather than as "awards." It happened that my own financial support was neither a Research Council studentship nor an assistantship, but a university scholarship, and after a while I noticed that I was receiving slightly less money even than the other students. I enquired about why this was so and was told by the university offices that the scholarship paid less because it carried "extra prestige." This was my first introduction to the great world of status. At the time I was surprised, but had I been longer in the tooth, I should have understood perfectly.

I had a good time in Edinburgh climbing many of the Scottish mountains and for a year editing a rather pretentious literary magazine. But the most interesting thing that happened in the lab was the curious disease suffered by one of the other students by the name of Simon Law. Simon was from Canada, and since he was two years ahead of me, I held him in great regard and assumed that he knew everything there was to be known about the subject. He was small, fairly quiet, and retiring, with a wispy beard. Nowadays a beard is a badge of belonging to the dim and distant past, rather like Karl Marx or Charles Darwin, but in the early 1970s, everyone had a beard, including me. One day Simon surprised us all by putting up on the notice board some enormous cartoons about ribosomes. Ribosomes are the tiny bodies within the cell that carry out the assembly of protein in accordance with the sequence of nucleotides in the messenger RNA. Despite their small size, ribosomes are very complex, and one ribosome consists of an assembly of about 80 molecules of protein together with 3 molecules of structural RNA. It was the ribosomal RNA that our lab was mainly working on. Simon then started talking to us, instead of skulking quietly in his room all day, which is what most scientists do when they get the chance. He talked about the work, he talked about how he wanted to reinvigorate our supervisor, he talked about politics. The next day he talked even more and surprised me by getting some potatoes out of a freezer and rolling them down the corridor to illustrate some point or other.

I went with him back to his flat to eat and asked him why he was so excited. He made oblique references to a woman. So that was it! He was a shy and retiring character, so this might be his first girl ever. No wonder he was excited. The day after that, he was talking about Life, the Universe, and Everything. He went all round the building talking to people he had never seen before. We started to avoid him. He went into town and spent large amounts of money. One day he vaguely said that he was going to "write to his friends" and came back with a packing case full of picture postcards. A day or two after this, we heard that Men in White Coats had been called out by his flatmates and that he had been taken away to an institution!

When I visited him there, my first impressions of the mental hospital did not inspire confidence. A vast Gothic horror of a building set in enormous and, as far as I could see, totally deserted grounds. It was hard to find the front entrance, but as I peered in through some of the ground-floor windows, I saw white-clad figures grimacing in ghastly postures. Eventually, I found the front door, beneath the tower with birds (or was it bats?) flapping in a sinister way high above. The reception desk was deserted. Gingerly, I crept through a vast dining room with stags' heads staring sternly down at me. I entered a subterranean passage and passed a door through which I could see a centrifuge and other equipment clearly intended for experimentation on the unfortunates incarcerated in this dreadful place. I was terrified. But as I penetrated still further into the labyrinth, things began to improve, and when I did finally find Simon's ward, it all seemed much more normal, just like any other hospital. Maybe all this stuff near the entrance was just to deter potential escapees.

I talked to Simon for a while, but he was fairly unresponsive and obviously doped up to the eyeballs. I also saw his doctor, who seemed interested in any information about the background to the crisis. "There was a woman," I said, still half believing that the disaster had really been precipitated by some tempestuous love affair. The doctor seemed uninterested in half-mythical females. "He used to write the labels on his bottles in Japanese characters," I said. "Oh, really?" said the doctor. "That's interesting. Tell me more about the labels." It was true enough. All scientists are obsessive about labelling everything, and most are obsessive about not wanting other people to use their solutions. Other people are usually slobs, bound to contaminate your solution with something that will ruin your next experiment, so Simon was pretty safe with his Japanese labels. I struggled to explain that this was actually quite normal behaviour but felt myself dropping my colleague deeper and deeper in the psychopathological soup with every sentence. Finally I escaped, feeling that my contribution had hindered rather than aided Simon's chances of recovery. He came out after a few weeks, and I found myself watching him like a hawk for the slightest sign of peculiar behaviour. It is surely true that you have a much narrower latitude of permissible normality if you have just had some psychiatric treatment than if you have never had any at all. Simon found it hard to return to work, and so in the end my supervisor took him to his half-built house for some occupational therapy. After a few months, he went back to Canada. His spectacular attack had, of course, been a textbook case of the manic phase of a manic-depressive psychosis.

At the time psychiatry was in the spotlight. Thomas Szasz wrote articles in *Nature* saying that there was no such thing as mental illness. R. D. Laing treated schizophrenics with psychotherapy and without drugs. *One Flew over the Cuckoo's Nest* was a massively successful film, showing a group of long-term mental patients who became quite normal when they escaped from their repressive mental hospital. It was the time when all medical students seemed to want to be psychiatrists, when the idea that there was any genetic component to any mental disorder was regarded as more or less undiluted Nazism. It was an interesting time to see a real psychotic illness at close quarters. I could certainly appreciate that people got locked up not because of the bizarre nature of their own perceptions but because of the trouble they caused for other people. This was, after all, what had happened to Simon. But I was really impressed by the magnitude and sudden onset of the disorder and by the way that it could be rapidly suppressed, if not permanently cured, by drug treatment. It seemed hard to believe that a little gentle psychotherapy could have been so effective.

In my own career, the results of the sink-or-swim process were not a great success, and I will not bore readers with the content of my Ph.D. thesis. It was not very interesting, mainly because molecular biology in those dim and distant days was still in the precloning era, when molecular biology projects like mine consisted mainly in grinding up cells or tissues, purifying substances from them, and measuring the percentage of this or that. As discussed in Chapter 3, *cloning,* or more precisely *molecular cloning,* means that the gene for the molecule you are studying is inserted into bacteria and can be grown to a very high copy number. This means that any gene or gene product can be prepared in large quantities. The practical difference that this has made is that the number of genes that can be studied has increased from the one or two that occur naturally at a very high copy number (including the dreaded ribosomal genes that made the ribosomal RNAs that we were studying) to the entire set of genes present in all animals, plants, and microorganisms. Nowadays any gene whatsoever can be isolated, analysed, manipulated, and its protein product prepared. Molecular cloning has enabled molecular biology to grow and flourish like no other branch of biology in the whole of previous history.

But as for myself, after two years I had had enough. I realised that the required standard for a British university Ph.D. thesis was not very high, so I decided to cut my losses and write up my notebook of uninteresting work there and then, thus giving myself some time to think about what to do next. I read. I thought. Should I move out of science? What on earth else was there that looked remotely interesting? Not a lot. No, I would stay in science and do something different from this ghastly molecular biology. I sat in the library for months, reading and thinking. Maybe virology? Maybe bacterial genetics? Eventually I settled on something that seemed almost as far away from what I had been doing as possible: experimental embryology.

Now this did look interesting. I was mainly influenced by a textbook called *Principles of Embryology,* written by C. H. Waddington in 1956. Waddington was actually still alive and a professor in Edinburgh, although by then he had lost interest in embryology and was trying to Save the Environment. In my humble position, I did not dare to go and talk to him, but I do remember going to a seminar at which he was present and regarding his every word with reverence. I remember being fascinated to find that there were many experiments that seemed pregnant with significance but for which there was no real explanation. For example, if a duck embryo, at the stage when it was a simple sheet of cells, was cut into two parts, two embryos would result. The rear half of the embryo retained the head-to-tail orientation of the original, while the front half had a random orientation. Or, if a limb rudiment was grafted into the flank of a salamander embryo between the fore and hind limbs of the host, it grew with reversed orientation; that is, its "thumb" was at the rear and its "little finger" at the front of the extra limb. In fact, there was a gigantic corpus of microsurgical experiments carried out mainly between 1920 and 1939 that pointed to a totally different picture of how animals developed from that provided by the molecular biology of the precloning era. Although I was too frightened to approach Waddington himself, I did talk to some of his assistants. One of them was actually employed to rewrite *Principles of Embryology* and bring it up to date. Alas, the Great Man eventually lost interest in this project as well, and his ghostwriter soon left to become a schoolteacher. It was not until 20 years later that I heard from him again. The book was finished! Of course it was not now called *Principles of Embryology* and Waddington's name was no longer featured, but I knew it was the same book. By then I was an advisory editor for Cambridge University Press, and it was with some pleasure that I was able to secure publication for this book, whose original version had launched me on my own career in embryology.

I should explain that although I now regarded (and still regard) myself as an embryologist, most workers in the field call themselves "developmental biologists." The general public has no idea what this is and usually thinks it must be something to do with improving crops for use in underdeveloped countries. The term is used partly because the subject includes various things that develop but are not embryos (regenerating organs, plant meristems, wound healing), but mainly because "embryology" sounds old-fashioned, particularly to Americans, and nobody in the hungry 1990s can afford to sound old-fashioned. An example of the power of this effect is provided by the scientific journal that used to be called the *Journal of Embryology and Experimental Morphology.* It had been started in the 1950s and had always been quite good for vertebrate embryology, but it lacked charisma and existed in quite a narrow scientific niche. In 1987 it changed its name to the aggressive, single word title *Development,* and the name change was accompanied by a larger format, more frequent publication, better pictures, and more hype generally. In no time at all, its impact factor went through the roof, and it came to dominate the field (see Chapter 4 for more about "impact factors"). Superstars who would previously only consider publishing in the fashion journals *Nature* or *Cell* started to send in their papers, and *Development* has never looked back.

The Middlesex

What finally tipped the balance for me was attending a lecture by the great Lewis Wolpert. He was, and is, well known to developmental biologists around the world for his immense enthusiasm and magnetic personality. He was by origin a South African and had started life as an engineer, become bored with it, and moved into academic biology. He came to Britain in the 1950s and did a Ph.D. at King's College, London, on the mechanics of cell division in *Amoeba.* He then spent a short time as a lecturer (assistant professor) at King's College, during which time he developed an interest in the theoretical issues of embryonic development. He experienced a meteoric career progression, and only six years after gaining his Ph.D., he had become a full professor and chairman of the department of biology at the Middlesex Hospital Medical School in London. He had the great advantage of not having been contaminated by the molecular biology of the day and so had a completely novel outlook on the problems of development. He formulated a new theory called the theory of positional information, which attempts to explain many of the bizarre results of classical experimental embryology and is now known to be substantially correct. The essence of the theory is that cells need to be told their position within the embryo before they can decide what sort of differentiated cell to become. They may be told their positions by various mechanisms, but one in particular that is associated with Lewis's name is a concentration gradient of an inducing factor, an issue to which we shall return in Chapter 5. At the time, Lewis was doing experiments with a tiny freshwater animal called Hydra, which is a tube a few millimetres long with a sticky foot at one end and some tentacles at the other. Actually, he never did any experiments himself but entrusted them to his assistant Amata Hornbruch, who was herself a very fine microsurgeon.

The remarkable regenerative abilities of Hydra had been discovered as long ago as the eighteenth century by the Swiss gentleman scientist Abraham Trembley, who had described their ability to regrow missing heads and feet. This work created a fashion for experimental biology that has perhaps never been equalled. The French philosopher Voltaire is alleged to have remarked at the time that the progress of science was so rapid that it was soon going to be possible for human beings to remove an unsatisfactory head and cause it to be replaced by a better one. The French Revolution, which followed soon after, certainly bore out the first part of his prediction, but the second part has so far remained immune even to the best and most expensive endeavours of modern molecular biology.

Lewis Wolpert happened to give a lecture about Hydra while I was in Edinburgh, and I was fascinated. It seemed that the formation of a new head or foot was all a matter of gradients of inducing factors that evoked different cellular behaviours at different concentrations. These factors were hypothetical, as all the experiments were very simple microsurgical ones involving chopping up and rearranging the tiny creatures. But it seemed amazing to me that one could really find out about mechanisms from such simple manipulations. It sounded a lot more fun than grinding things up and measuring percentages. So I went to his lab as a postdoctoral fellow. I was still too naive to ask myself whether this was a "good place" and whether this was a smart career move or an excursion to nowhere. As it turned out, his lab was a very good place, if that means being really stimulating for its members. It would, perhaps, not have been regarded as such from the outside, because at the time the things we were working on were almost unknown to the wider scientific community, and it was often quite difficult to explain to other biologists what we were trying to do.

The Mysterious Bob Franklin

One of the pots of money I applied for to enable me to go to Lewis Wolpert's lab was a most curious type of fellowship that never quite came into existence. University notice boards in 1974 sprouted some mysterious announcements about "Bob Franklin Fellowships." Nobody seemed to know anything about this scheme or about who Bob Franklin might be. But they looked pretty good: five years of support and no particular restrictions about how to spend the money. Candidates could not apply themselves; they had to be nominated. So I asked our head of department to nominate me and, as things like this cost nothing, he did. Actually, a couple of weeks later, the head of department was away, and I wasn't quite sure whether he had sent in the nomination, so I also asked my supervisor to nominate me. As it cost him just as little, he did too. I was quite pleased to be invited to the interviews in London a few weeks later. These took place at a hotel in London. All expenses were to be paid by Mr. Franklin. No details about the selection process were provided; "all would be revealed" in good time. So I turned up at the hotel, and once again there were no details about what was to happen. We were just told to gather in a meeting room at 9 A.M. the next morning. By this time, I was having serious doubts about the scheme, and in particular about Mr. Franklin, whom I suspected of not existing at all.

However, the next morning I was proved wrong, as Mr. Franklin appeared in person. There were about 40 people there for interview and five fellowships on offer. Bob Franklin was an Australian, tall and slim with greying hair. He had the decisive air of one who has been successful in business, and we learned that he had made a large pile through international real estate deals. He started out by telling us that we were the crème de la crème, the very best biological scientists from the whole country. The proof of this was that every one in the room had been specially nominated by some eminent scientist or other. Some, he said with a flourish, had even been nominated twice! I squirmed partly at the thought of the slender basis of my own nominations and also because at this early stage in my career, I thought the flattery rather odd, when it was so obviously based on nothing.

Mr. Franklin then asked, as a preliminary warm-up, for everyone in the room to stand up, say who he was, where he came from, what he was working on, and whether he believed in God. This was a surprise, or perhaps not such a surprise, as we all knew there was something very odd about the whole business, and for Mr. Franklin to be a religious nutcase seemed quite plausible. The opening introductions, however, took rather longer than expected. The first few people literally just stood up, answered the questions, then sat down. But as we went round the room, the answers began to become longer. The longer they became, everyone reasoned, the less time there will be for anything else, so this introduction itself must be a vital part of the selection! By the time we were half way round, the brief statements describing the candidates' research projects had reached about 15 minutes, and the last person went on for half an hour. Nobody spent too long on the religious part of the question, but there was a definite tendency to expand the spiritual side of the personality. The few out-and-out atheists such as myself couldn't bring ourselves to say that we believed in God but definitely felt that we were doing ourselves down. As the introductions had all taken so long, Mr. Franklin then announced that we would now break for lunch, and he disappeared.

Tongues wagged excitedly over lunch. What was this all about? Why did he want to know about people's religion? My own theory was that he wanted to hire some ace biologists to find the molecular basis of the soul. But again I was proved wrong with alarming speed. No sooner had I propounded my theory than one of our number burst in excitedly. He had, it seemed, been granted a personal interview with Mr. Franklin. He had been told that he was an individual of supreme ability, and undoubtedly the best candidate, but that he would be unable to receive a fellowship as he was a Roman Catholic. Well, this was a surprise! We had all assumed that a positive answer had been required to the God question, but now it seemed that all these people stretching their belief had not done themselves any favours. In the hungry 1990s, there is no doubt that the atheists would have given a whoop of joy, booted out the believers, and claimed the loot. However, this was not the 1990s; it was the 1970s, just a few years after the main period of student revolt in universities around the world. Many of those finishing their Ph.D.s in 1974 had been student activists in 1968 and 1970. They still had a very strong sense of idealism and a strong opposition to any form of prejudice or discrimination. Because of this, a wave of disapproval swirled round the room, and when Mr. Franklin returned a few minutes later, the mood had turned rather ugly.

He now faced a very hostile audience. "What was this discrimination for?" "Why did it matter what religion someone held?" Mr. Franklin was not used to this sort of thing. He had the money and expected those who wanted his money to allow him to determine the rules. He would not explain except to say that his long experience had persuaded him that religious belief was not compatible with creative work in science. He tried to quieten the audience and to move on to the first stage of selection. In fact, it seemed that he had already made the selection, largely on the basis of the religious question, and wanted to read out the names. "No!" we shouted, "No names!" We knew that as soon as the names were read out, the unity of the crowd would be lost and the unselected would have to leave.

In fact, the names never were read out. The discussion went on all afternoon, but there was to be no compromise on the vexed issue of the religious discrimination. So the scheme collapsed. We all dined at Mr. Franklin's expense that evening and then went home to dine out many times more on the extraordinary tale. Today it seems quite extraordinary that some compromise could not have been found, that such a large amount of money could have been thrown away so impulsively. Trying to rationalise it afterwards, I felt that perhaps it was a good thing, as Mr. Franklin was obviously not the sort of man who could have left his research fellows alone. Had they been appointed, he would undoubtedly have been pestering them and interfering in their research and eventually make the whole scheme unworkable. Still, I also ruefully remember the remark of one of the candidates, now a Fellow of the Royal Society and a very senior figure in British science. He also was an atheist, and in the midst of the row, he turned to one of the believers and said, "I do hope that you will remember this and do the same for us when we are down below, roasting on the griddle." Time will tell.

How Many Fingers?

Actually, it didn't matter too much about Bob Franklin, as I got another fellowship to go to Wolpert's lab. When I got there, I found a neglected colony of axolotls and immediately fell in love with these magnificent animals. The axolotl is a type of salamander that lives all its life in water and never metamorphoses: the adults retain their gills and fleshy tails, but they also have legs. As their name suggests, axolotls have their origin in Mexico. They used to be abundant in Lake Xohimilcho, which lies on the outskirts of Mexico City, but the city has grown so fast that there is no longer much left of the lake and not many axolotls left in the wild. Perhaps this is no bad thing, because the Mexicans are alleged to eat them. Fortunately for the axolotl, it endeared itself to Western scientists back in 1863 when 34 specimens were shipped to the Museum of Natural History in Paris. Observers were amazed when these apparent larval forms were able to breed, since salamanders normally reach sexual maturity only after metamorphosis. They were even more amazed by the subsequent spontaneous transformation of some of the animals into the postmetamorphic form, resembling adult salamanders of other species. It was later found that metamorphosis can be induced by treatment with the hormone thyroxine, or with iodine, an essential component of thyroxine. Axolotls are easy to rear in captivity, and the original group of 34 has expanded into numerous laboratory colonies all over the world.

Anyone who keeps axolotls will know that it is almost impossible to stop the young ones nipping off each others' legs. This may sound distressing, but in fact, the animals hardly seem to notice, as they do not need their legs for anything much and the legs themselves grow again within a few weeks. This remarkable ability to regenerate missing structures has always fascinated those who know about it. Almost every grant application to work on regeneration carries the covert, and often the overt, message that this is the route to human regeneration. If only this application were supported, we would learn how the axolotl does it, and if only we knew how the axolotl does it, then maybe we could do it too. So for many years, the axolotl's main role as a laboratory animal has been for the study of limb regeneration, although they are also used to some extent for studies of early embryonic development, the development of neural pathways, and metamorphosis.

I decided that I would use the obliging axolotls to have a crack at one of the enigmas found in Waddington's book. Back in 1933 a Ukrainian biologist called B. I. Balinsky, later to become famous for writing a standard textbook of embryology, published a paper describing the induction of extra limbs in the flank of newt embryos. This was achieved by the implantation into the flank of tissue from the developing nose, a structure called the nasal placode (Fig. 2.1). The induction of extra limbs seemed remarkable enough, but even more remarkable, instead of being normally oriented with the thumb at the front and the little finger at the back, most of them were the other way round, with the little finger at the front and the thumb at the back. By this time Wolpert's group had moved on from Hydra to limb development, and they were working mainly with chick embryos. They had concentrated their attentions on a small region of tissue at the posterior edge (i.e., the future little finger edge) of the limb bud, which seemed to emit a gradient of inducing factor controlling the pattern of digits that subsequently developed from the remainder of the bud (Fig. 2.2). If a similar signalling region existed in the newt, I reasoned, this would explain the reverse orientation of extra limbs that had been caused to form in the flank, since they would lie behind the source of the signal rather than in front of it.

I persuaded the axolotls to lay some eggs and attempted to implant pieces of nasal placode into the flank. Not a single induction! I tried again, but it just didn't work. By then I was six months into my two-year fellowship, and I decided that I needed some results. So I grafted some bits of the limb rudiment itself into the flank, since if these didn't grow then nothing

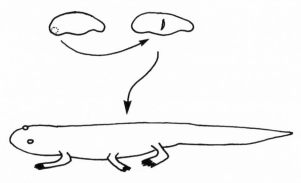

Figure 2.1. Balinsky's experiment: induction of a limb by a nasal placode.

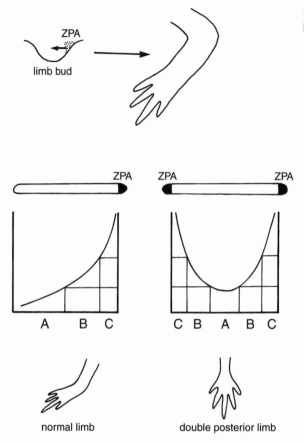

Figure 2.2. The zone of polarizing activity (zpa).

would, and I could give up. Lo and behold, they did grow, and not as normal limbs but as mirror duplications. This means that the resulting limbs had a symmetrical pattern with, for example, a little finger, a fourth finger, a third finger, another fourth finger, and another little finger. This pattern of structures was immediately recognisable to anyone in Wolpert's lab because of the concurrent work with the chicks in which such duplications were regularly produced. I decided that this result could be explained if the polarizing activity in the axolotl embryo lay in an extended region of the flank, such that any graft would get a signal from both front and back edges. Once off the ground with this group of experiments, I spent most of the rest of my time at the Middlesex filling in the idea with more data.

Some time later I bumped into Professor Newth from Glasgow University. He had been Lewis's predecessor at the Middlesex, and for all I know, they had originally been his axolotls. I knew that he had repeated Balinsky's induction experiments in the 1960s, so I asked him why it didn't work for me. "What species were you using?" he asked. "Axolotls," I said.

"It doesn't work in axolotls," he said. "It only works in newts." In science glory or disaster is born from such chance.

Brussels

While I was at the Middlesex, I signed up for an advanced course in developmental biology at the Free University of Brussels. This was quite a laid-back course, extending over three weeks, and included practical classes on a wide variety of creatures ranging from mouse embryos to slime moulds, as well as a number of lectures. Most of this is now rather hazy in my memory, although I remember clearly that we had to scour the restaurants of the city each evening to find something to eat consistent with our meagre budgets. There is a Flemish dish called *stumph,* which is made from potatoes, onions, and cabbage. Its appearance is well conveyed by the name and fitted our financial requirements quite well. Fortunately, even in expensive Brussels, beer was still cheaper than in Britain.

Anyway, at this course I was privileged to meet two of the old timers of embryology: Jean Brachet, who was running the course, and Tuneo Yamada, who gave one of the lectures. Both had worked in the fabled "elder days" of experimental embryology between the wars (that's between World Wars I and II, I have to remind my young colleagues, who were still oocytes at the time of Vietnam). Brachet was tall, genial, and avuncular and seemed very French, although, like Hercule Poirot and Georges Simenon, he was one of those Frenchmen who is really a Belgian. His studies on nucleic acids in the 1940s had helped to lay the basis for the later explosive growth of molecular biology, and because of this, he had considerable influence in Belgian science. He had succeeded in persuading the Free University of Brussels to build him a considerable building for his developmental biology institute, and because this was Belgium, he did not just have one building but two: a larger French-speaking institute and a smaller Flemish-speaking one. He loved to reminisce to the callow youths and maidens taking the course, and as this was invariably accompanied by clouds of smoke from his pipe and ghastly noises from his repeated throat-clearing, it made a lasting impression on them.

Tuneo Yamada was quite different from Brachet. He was very dapper, quiet, reserved, polite, and, like many Japanese, looked about 30 years younger than his true age. Later, when I was able to talk to him at length, I discovered that the first Japanese scientist to launch serious embryological research had been a Professor Ishikawa, who had specialised in fish development. He had introduced Yamada to the subject, and Yamada had duly gone to Germany. In the 1930s Germany was the world centre for embryological research. The journal called *Wilhelm Rouxs Archives für Entwicklungsmechanik des Organismen* held the position that *Development* holds today. Anyone who was serious had to go and work in Germany for a period, and if really serious, had to publish his papers in German as well. In Germany Yamada had done his pioneering work on how the various tissue types arose from the mesodermal cell layer, about which more below. When World War II started, he was drafted into the Japanese army and spent five years on a Pacific island. Fortunately for him, the Americans left his island alone and he survived the war. By 1946 he had managed to get back into science in Japan and published a number of further works about the early amphibian embryo. In 1961 he moved to the United States and took up the problem of lens regeneration in the newt, for not only can newts regenerate limbs, they are also able to reform the lens of the eye after its removal. The new lens grows not from the skin, from which the original lens was formed, but instead from one edge of the iris. He moved to Switzerland in 1974 and continued to work on lens regeneration until his death in 1997.

Brachet introduced Yamada at some length, and although the lecture was on lens regeneration, Brachet looked long and hard at me and said, in his inimitable Franco-Belgian accent: "And for zose of you 'oo are interested in gradients, may I remind you zat Yamada 'as discovered a gradient in ze mesoderm. Ze dorsal part forms notochord, ze ventral part forms ze blood islands, and ven you put ze two together, you get ze muscle."

So, there was a gradient in the mesoderm! I was interested in gradients because I was in Lewis Wolpert's lab, and we knew that there were gradients of inducing factors in Hydra and in the limb. At its simplest, a gradient meant a concentration gradient of a substance to which cells would respond in different ways at different concentrations. The gradient is emitted from a special region of cells producing the substance: a signalling centre or organising region. Since the surrounding cells can respond in different ways at different concentrations, a whole series of territories can be formed in response to a single graded signal. When an extra signalling centre is grafted into an embryo, in a different place from that in which it is normally found, an extra set of territories will be induced radiating out from the graft (Fig. 2.2). I went and looked it up in Waddington's book, and sure enough, there was a whole page about Yamada's experiments. I decided that this was what I wanted to do. I would work on the early amphibian embryo and apply the Wolpertian theory of gradients to the early stages of development of the whole body. In other words, I would try to understand how the early population of cells that are formed by the division of the egg becomes divided into different territories, each of which is committed to form one of the major structures of the body.

Richmond Hill

My fellowship at the Middlesex was only for two years, so I had to find a job. I was rejected for various lectureships, fortunately as it turned out, for the 1980s were a period of unrelieved gloom in British universities, and to keep a research programme afloat amid all the reorganisations, cuts, demands for extra teaching, market orientation, appraisal, audit, and all the rest of it demanded a degree of superhuman ability that I definitely do not possess. Fortunately for me, a small and little known institute belonging to the Serious Disease Society (SDS) had just undergone a reorganisation and was looking for scientists. This institute was at Richmond Hill in South London. The SDS had recently moved most of its labs to a large and lavishly appointed institute in Central London and had wondered for a while what to do with the rump left at Richmond Hill. In time-honoured management style, they had eventually appointed a new director, given him some money, and left him to sort things out.

Like many incoming directors, our man wielded the new broom with vigour, and the lab heads remaining in the institute either left or were expelled to remote outhouses. He had wanted to appoint two new principal lab heads in addition to himself, but one of these, the American molecular immunologist James Samson, was a real prima donna and caused no end of trouble. One month he was coming, and the next he wasn't. One week he agreed to the terms, the next week he wanted more money or more space or more facilities, a week later he agreed again. In the end he never did come. But by then the SDS had fitted out the whole of the top floor for him and purchased a great deal of equipment. One day, when it was quite certain that he would not come, the word was given and it was like Christmas time! The whole institute staff had carte blanche to go and open Samson's boxes and take what they wanted for their own labs. The Samson debacle meant that the place was quite well appointed and well equipped but needed some more bodies to fill it up. This was achieved partly with new junior appointments, like myself, and partly by existing junior scientists taking over the empty space and becoming lab heads themselves by right of conquest.

So, thanks to James Samson, whom I had never met, I had got into the Serious Disease Society. Developmental biology has always had a small corner of medical research because serious diseases often depend on interactions between cells, and developmental biologists are the people who mainly study such interactions. So, although my work was on the fringes of the SDS's activities, it was not entirely ridiculous to be interested in such things as limb development or how the animal body plan was formed from a featureless egg.

My first problem was that for the first four months, all my embryos sickened and died. Now that I was on my own, and not someone else's student or postdoc, I found that nobody else cared a jot about my embryos. As far as they were concerned, they were very busy people and my embryos would have to sort out their own problems. I knew I should have to sort this problem out before I could do any work. Our director was very supportive, although he had no direct knowledge of embryos and could not really give practical help. After a long investigation, I finally tracked down my problem to the water. Yes, the water! It is every scientist's nightmare to have an unreliable water supply because water is the one ingredient in every solution that you take for granted. But all the distilled water in the institute came from a single distillation plant in the washing-up room containing a large, bare copper element. It seems to be well known that enough copper can get across in the spray in such stills seriously to contaminate the distillate. Once this was discovered, I bought a new still with its element covered by glass, and over the next few weeks so did everyone else in the institute. Even though they had not themselves noticed any problems with the water, the mere thought of contaminated water was too much for them to bear.

I continued my work on limb development and regeneration for a few years while at the same time working up a project to study gradients in the early embryo. For this new line of work, axolotls were not so suitable, and I moved over to the famous African clawed frog *Xenopus* (pronounced "zenopus"), which are a very good experimental material. We shall meet *Xenopus* at greater length in Chapter 5.

Oxford

So I gradually got things going and started on the line of work that would, 10 years later, culminate in the experiment described in the first chapter. By that time my lab had moved to a unit in Oxford, still funded by the Serious Disease Society. Oxford is a great university, but one of the things remembered with fondness by the residents of the city has nothing at all to do with the university, and it occurs in September each year, when the students are not in residence. This is the St. Giles Fair, a traditional travelling fair that occupies for two days the whole extent of St. Giles, which is a very wide historic street near the city centre. As it happens, Oxford residents do not realise that the fair is not just for them; it actually travels around the whole south of England. In some towns, such as neighbouring Thame, it takes over the city centre for a whole week. But this fact is little known to the English intelligentsia, as Oxford is fashionable and the other towns are not.

Shortly after arriving in Oxford, I was keen to see this fabled event and duly went along to St. Giles. The fair was crowded and noisy, with a strong smell of fried onions and candy floss. By this stage in my life, I had lost any appetite for being whirled around on those large centrifuges that seemed more suitable for spacemen than scientists. But the teenagers of Oxford had no such inhibitions and pressed with abandon onto the machines to have their stomach contents churned at considerable expense. My interest was entirely different, for I had heard that the St. Giles Fair was the last refuge of the freak show. It is hard to imagine anything less politically correct than a freak show, but in the incorrect days of the 1930s or even the 1950s, it was not uncommon to find at travelling fairs a stall exhibiting the "Fattest Lady in the World" or the "Tallest Man in the World." My own perception of freak shows was entirely conditioned by the famous movie *The Elephant Man,* in which an individual of truly repulsive appearance was discovered by the famous surgeon Frederick Treves and brought to the London Hospital for investigation. The Elephant Man himself, the so-called "disgusting specimen," was called Joseph Merrick and turned out to be a gentle man of high sensitivity. The exact nature of his condition is unknown to this day, but it may have been neurofibromatosis, a disease involving formation of multiple benign tumours from the cells of the nerve sheaths. Certainly, he looked quite spectacular in the movie, with projecting tumours all over his body, a massively overgrown cranium, and a face consisting mainly of an elephant-style proboscis. On account of his unfortunate appearance, he used to wear a sack over his head in public with just a small slit for his eyes. Treves was unable to cure him but used his influence to allow him to live out his days in relative comfort in the basement of the hospital. His original motivation had been scientific investigation, but he had suffered some pangs of conscience when he realised that exhibiting Merrick to meetings of pathologists was not so very different from leaving him in the fairground stall to be exhibited to the general public.

Anyway, I searched for the freak stall feeling sure that a developmental biologist like myself could, if not actually cure a fairground freak, at least find something interesting to say about him or her. I was delighted to find a stall advertising "The Tiniest Man in the World" and paid over my pennies to go in. The tent was arranged with an outer antechamber, bathed in dark red light to achieve the maximum of suspense and anticipation. As I slowly walked through, the tension mounted in my breast. What would I find? How small would he really be? Would he be in pain? On the walls were pictures of all manner of human freaks including dwarfs, giants, and many pairs of Siamese twins. This was obviously a place in the finest traditions of human freak shows! Inside was a raised walkway surrounding a well in which sat "Wee Willie," the Tiniest Man in the World. According to the large and gaudy placards outside the tent, he was only 26 inches tall. But I have my doubts. It is well known among psychologists that our perception of deviations from the human norm is always rather magnified. So a man, say, four feet high appears as a tiny dwarf even though he is 70 percent of average height. This effect was heightened by the fact that we were looking down on Wee Willie from the raised walkway. As his name suggested, Willie was a Scotsman, and a very garrulous one indeed. He talked incessantly, which had the positive effect of rapidly breaking down the barrier between himself, the freak, and us, the fee-paying audience. So we chatted for a bit and he sold me a copy of his photo, which I still have to this day. Then I left, feeling that, unlike the Elephant Man, he probably would not want to be rescued and accommodated for the rest of his life in the basement of the University Zoology Department. I did not dare ask him how tall he really was. He looks very small on the photo, which shows him sitting on the front of an enormous American car, his legs pointed toward the camera to foreshorten the image. He was certainly fairly small; my guess would be about 3 feet 6 inches. He was of course an achondroplastic dwarf, quite a familiar abnormality in most parts of the world, and a member of a tribe that can sometimes make a good living from appearing in movies or stage productions that have a need for dwarves, such as *The Wizard of Oz* or *Charlie and the Chocolate Factory.*

Achondroplasia is a heritable condition. It is dominant, meaning that 50 percent of the offspring of an affected individual will themselves show the trait. The actual defect is a failure of the long bones of the limbs to elongate sufficiently. Normally, there is a protracted period of bone growth from birth to puberty. During this period a "growth zone" composed of cartilage persists between the ossified shaft and the ossified terminus of a long bone and continues to add new material to the shaft. The cells of the growth zone continue to divide until at puberty the zone shuts down and the ossified regions on either side finally fuse together. Although I had not really been expecting to be able to cure Wee Willie, by a curious coincidence it turns out that I could almost have done so. By this time, immediately after the experiment described in Chapter 1, I was immersed in the properties of the fibroblast growth factors (FGFs), and a few years later it was discovered (not by us) that achondroplasia is actually caused by a mutation in one of the receptors for the FGFs. The effect of the mutation is to make the receptor signal continuously even if there is no FGF present to stimulate it. This means that in normal circumstances, the shutting down of the growth zone is controlled by FGF signalling, and in the case of the achondroplastic individual, the shutdown occurs prematurely because the receptor is active all the time. It is quite possible that a specific inhibitor of this FGF receptor could be devised, and if this were administered to achondroplastics during the phase of long bone elongation, the shutting down of the growth zone could be delayed until they had reached a more normal height. So I could not cure Wee Willie in 1986, but I did later gain some satisfaction from realising that I was working in exactly the branch of developmental biology that would one day lead to a treatment for others with his condition.

The unit in Oxford did very well in conventional scientific terms, for it contained "good people" who published their results in "good" journals. In fact, the unit participated in a significant way in the international boom in developmental biology that occurred in the 1980s. But good things do not last forever, and the unit itself had a rather short life of 11 years. The reasons were entirely financial. When I joined the SDS, it had enough reserves to run for at least four years with no income. Seventeen years later, the reserves had shrunk to a point that there was only enough to run the organisation for three months. Since the income consists mainly of legacies, which are by their nature totally unpredictable, something had to be done. As the cold air of reality entered the organisation, the cuts began. Over a few years, 12 lab heads and several other staff were made redundant or forced to retire early. Despite its scientific success, our own unit became scheduled for closure, which meant either leaving or moving to one of the London labs with its associated commuting. As I didn't fancy commuting for the next 20 years, I felt that the time had come to leave. Fortunately, my 20 years of full-time research had made me a reasonably attractive candidate for university jobs, and I am now a professor at the University of Bath, a small but successful science-based institution in the South West of England.

The Growth Factors

If you are an eco-activist, genetic engineering is one of those disasters that has recently afflicted mankind and caused suffering to millions on about the same scale as Genghis Khan, Hitler, Stalin, and the Four Horsemen of the Apocalypse. If you are a financial investor, genetic engineering, in its guise as "biotechnology," is a label for stocks that can show spectacular growth, preferably when the quoted company is taken over by one of the big pharmaceutical houses. If you are a biological scientist, genetic engineering has transformed your academic world and promises to transform much of society over the coming decades.

Genetic engineering really means making animals, plants, or microorganisms with an altered genetic constitution. This has been done by humanity since the dawn of agriculture using traditional selective breeding methods, and very profound changes have been brought about in crop plants and in domestic animals. However, what you can do by selective breeding is limited by the amount of genetic variation already present in a population. You can select for more or less of gene variants that are already there, but selection alone cannot create new genes or new gene products. By contrast, modern genetic engineering can, in principle, create new genes and introduce them into any organism at will. This has created some anxiety among the general public, although the fears are probably exaggerated and the prospects for beneficial applications are extensive.

Although it became possible to think about genetic engineering as soon as the structure of DNA was discovered in 1953, its practical implementation has depended entirely on the techniques of molecular cloning introduced in the 1970s. These techniques have enabled genes to be isolated and manipulated, and enormous numbers of novel proteins to be discovered, or if previously known, to be prepared for the first time in useful quantities. This ability has brought about phenomenal changes to most areas of biology. Complex processes—such as those of development, immunity, or cancer—that used to be understood only at the level of whole organisms are now understood also at the level of molecules. The technology has not only brought a new level of understanding, it has also opened the road to the creation of a whole new industry using molecular biology to make new products and new organisms. These dramatic changes make it appropriate to define 1973 as the "year zero" of the molecular biology calendar, as this was the year of publication of the first paper describing a method of molecular cloning. The history of biology can therefore be divided into the BC (Before Cloning) and the AC (After Cloning) era. Among the areas of biology that have exploded in the AC era is the study of growth factors, although, as we shall see, the bases for this subject were laid during the prehistoric BC times. To read the following narrative, and some sections in the remainder of the book, requires a little knowledge of molecular biology jargon. For the benefit of those readers for whom things like *sequences* or *libraries* are not features of everyday life, the essential background is briefly summarised in the postscript to this chapter.

The term *clone* has become a household word, but it can create confusion because it is used to mean several different things. Biologically, a clone is any set of perfect copies of something arising by a process of asexual reproduction. These may be DNA molecules, cells, or whole animals or plants. When we speak of "cloning a gene," we are referring to molecular cloning: the isolation of a DNA molecule coding for a particular protein and its amplification to a usable quantity by growth in bacterial cells. When we speak of the possibility of cloning a human being, we mean the replacement of the DNA in the egg with that from a single cell of an existing person, and the resultant growth of an exact genetic copy of the person. Although some animals have been cloned, human cloning does not exist, and some people think that it should never be permitted. Molecular cloning on the other hand now occurs every day in biological laboratories around the world, and the techniques of molecular cloning are studied by every biology major.

Growth Factors

Growth factors will undoubtedly be the magic potions of the twenty-first century. They are already invested with the same sort of mythology that graced hormones in the early years of the century and vitamins in the 1930s and 1940s. They are proteins, present in animal tissues at very low concentrations, having very high biological activity, and they are responsible for controlling some of the most essential of biological functions of cells, such as growth, differentiation, and survival. Those that were discovered by immunologists tend to be called *cytokines* rather than *growth factors,* but they are not fundamentally different from the rest. Growth factors differ from hormones in that they are not usually present as circulating agents in the blood stream but are found only in particular places, normally near the cells that produce them. This localisation often arises because they are bound to cell surfaces or to extracellular materials (the *extracellular matrix*). The behaviour of a particular cell, in terms of growth, differentiation, and survival, depends on its immediate environment, and the repertoire of growth factors found in this microenvironment is the most important controlling factor. Since growth factors are themselves produced by cells, they are the key to understanding the way in which different cell types in the same tissue affect each other's behaviour.

There are probably about 200 different growth factors known at present. The variety is considerable, but there are certain common features. They are all proteins, and this is another regard in which they differ from the more familiar hormones, some of which are small molecules such as steroids. They are all secreted from cells. They all exert their action by binding to specific receptors on the surface of the target cells. These receptors produce metabolic changes within the target cell, which eventually activate or repress specific genes to bring about a change in cellular behaviour. Many growth factors have been discovered by cancer researchers in the belief that they were investigating some specific feature of cancer cells. But in all cases the factors themselves have eventually turned out to be normal ones, essential for the normal development and functioning of the body. But derangements of the factors, or their receptors, intracellular signal transduction mechanisms, or target genes, are indeed often found in cancer.

Nerve Growth Factor (NGF)

By a curious paradox, the first growth factor to be isolated is not really a growth factor at all, in the sense that it does not stimulate cell division.

But it does cause extensive sprouting and elongation of certain sorts of nerve cell, namely, the sensory and sympathetic neurons. As is often the case, the story started with a cancer research problem. It had been shown in the 1940s that implantation of a particular type of mouse tumour into the chick embryo would attract the growth of sensory fibres from the nervous system of the host. This was further investigated by an Italian scientist, Rita Levi-Montalcini, who was a visiting worker in the lab of Viktor Hamburger at Washington University, St. Louis, Missouri. Hamburger was originally a student of the great Hans Spemann (see Chapter 5) and had moved to the United States when the Nazis came to power in Germany. Levi-Montalcini had graduated in medicine at the University of Turin and after World War II moved to the United States to work with Hamburger as a research associate. Hamburger had earlier shown that the removal of a limb bud from a chick embryo would cause extensive death of neurons in the part of the spinal cord that normally supplied the limbs. From these results, he had proposed a theory stating that the survival of these neurons depended on the target tissue. The idea was that some of the neurons would send their axons into the limb and that these particular cells would survive because they could absorb a protective factor from the tissues of the limb. The factor was transported back down the nerve axons to the nerve cell bodies and there prevented the cell death that was otherwise preprogrammed into the population. The neighbouring neurons, which sent axons elsewhere, did not receive the factor and thereby died.

Levi-Montalcini quickly decided that this was a suitable problem for investigation using the then arcane techniques of tissue culture, meaning the growth of cells or organ cultures in tubes or bottles, outside the body. Nowadays we can all do tissue culture in our own labs, but in the early 1950s, it was a difficult technique understood only by a few experts. Unable to do tissue culture herself, she arranged a collaboration with a group in Rio de Janeiro and soon flew down to Brazil carrying in her handbag the mice bearing the transplantable tumours. In Rio she managed to show that an isolated sympathetic ganglion from a chick embryo would send out masses of nerve fibres if it was cultured in a dish next to an explant of the tumour. So it seemed clear that the tumour was emitting a diffusible substance promoting the growth of the neurons. She called the activity *nerve growth factor,* or NGF.

Purification of the factor was undertaken back in St. Louis by a biochemist, Stanley Cohen. He had done his Ph.D. in biochemistry at the University of Michigan and was by this time a research fellow at Washington University. To begin with, it was not known what type of substance possessed the NGF activity, so at an early stage in the work, he treated the active tumour extracts with snake venom to see whether a nucleic acid was the active principle. Snake venom was known at the time to be a rich source of nucleases, the enzymes that degrade nucleic acids. Remarkably, the venom not only failed to destroy the NGF activity but instead greatly enhanced it. Snake venom was in fact itself a very rich source of NGF, and it was soon discovered that the factor was a protein and not a nucleic acid. Cohen then explored other mammalian tissues to find alternative sources of NGF. He soon found that the salivary glands of the mouse, which can be regarded as the evolutionary homologues of the snake venom glands, were also a very rich source of the factor. Using these two unusually abundant sources, he was able to purify NGF by 1960 and showed that minute quantities injected into chick embryos would provoke massive overgrowth of nerve fibres in the same way as the original tumour implants. This early discovery of similar factors from snake and mouse, both of which worked in chick embryos, should have alerted the scientific community to the similarity of developmental mechanisms between different sorts of animal. But there has always been great resistance to this idea, and even in the 1990s it can still be quite hard to persuade medically oriented researchers that the lower vertebrates have anything much to offer. Subsequent work on NGF showed that it was absorbed by the nerve termini and transmitted back along the axon to maintain the cell bodies. This work has amply confirmed the original theory of Hamburger that the target tissues produce factors that promote the survival of the central neurons to which they are connected.

Epidermal Growth Factor (EGF)

In the course of Cohen's work on NGF, he had injected some salivary gland extracts into newborn mice. Remarkably, he found that they caused early opening of the eyes and early eruption of the incisor teeth, effects that seemed to have nothing to do with nerve growth. Both of the effects were soon found to be due to a generalised stimulation of epidermal, that is, skin, growth in the newborn mouse. It became clear that this was not, in fact, an effect of NGF at all, but that there was another factor in the extracts, which became known as *epidermal growth factor,* or EGF. As with NGF, EGF is also present in quite large amounts in the salivary glands of mice, and purification of the protein was achieved by 1962. Using radioactive EGF, Cohen's lab, now at Vanderbilt University, showed that the target cells that were able to respond to the factor bore specific high affinity receptors on their surfaces. It was shown that the binding of EGF to cell surface membranes, isolated away from the cell contents, could provoke a phosphorylation reaction in the membranes themselves. Phosphorylation,

or more specifically, the addition of a phosphate group to a protein molecule, is now known to be a very important process in biochemistry. This is
because many proteins exist in active and inactive forms that can be inter-
converted by phosphorylation; thus, phosphorylation is the key step reg-
ulating their activity. The enzymes that add phosphate groups are called *kinases,* and those that remove them are called *phosphatases.* Could it be that
the EGF receptor was itself a kinase?

Initially, the EGF receptor could be studied only indirectly, by its ability
to bind radioactive EGF. Not until the 1980s did the advances in protein
purification technology make it possible to purify the EGF receptor, and it
was found that the same molecule did in fact both bind to EGF and also
act as the kinase. The receptor was a transmembrane molecule; in other
words, one end of the protein projected into the cytoplasm of the cell while
the other protruded through the membrane into the exterior medium. The
exterior domain carried the EGF binding site, and the cytoplasmic domain
carried the kinase site. Binding of the EGF to the exterior domain causes a
change in the shape of the protein, which is transmitted through the cell
membrane and activates the kinase. The phosphorylation of substrates inside the cell then initiates the cascade of further phosphorylations and
other interactions that eventually culminates in the activation of the genes
that bring about cell division.

A similar *modus operandi* of has turned out, in general outline, to be
common to all growth factors (Fig. 3.1). Cells are only competent to respond
to a factor if they carry the appropriate specific receptors on their surfaces.
Receptors for different growth factors do differ to some extent from each
other; for example, some of them are not kinases, although a large number
are. But all receptors are transmembrane molecules and have an extracellular part that binds the factor and an intracellular part that can initiate a

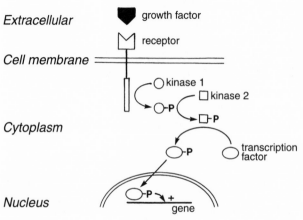

Figure 3.1. Mode of action of a typical growth factor. Only the phosphorylated form of each enzyme is able to phosphorylate the next one in the chain (P = phosphate group).

cascade of reactions leading eventually to specific gene expression. The metabolic steps between the receptor and the genes are called the signal transduction pathway, and very often involve a sequence of phosphorylation reactions between different intracellular kinases.

The purification of both NGF and EGF was achieved at an early stage because both are exceptionally abundant in the salivary glands that were identified as the best starting material. They are not nearly as abundant in other tissues, and in fact, they are not normally released from the salivary gland; otherwise, the entire body would be seriously overstimulated. Indeed, the reason for the presence of such large quantities in the salivary glands remains unknown to this day. Despite the relative abundance, the small quantities of purified protein available and the difficulties of protein sequencing at the time meant that the amino acid sequences of the two factors were not known until the early 1970s.

Other growth factors occur at much lower concentrations more nearly appropriate to physiological needs. They were correspondingly much harder to purify, and the majority were not properly characterized until the 1980s when it became possible to clone the genes and make large quantities of the proteins in bacteria or tissue culture cells. Moreover, the determination of amino acid sequences of growth factors became much easier at this time because of the introduction of rapid methods for sequencing the cloned DNA. So there was about a 20-year lag period between the purification of NGF and EGF and the real boom in growth factors. Once the boom started, the importance of the early phase became clear to everyone. Because they began it all, the pioneering work of Rita Levi-Montalcini and Stanley Cohen was recognised with the joint award of the Nobel Prize for Physiology in 1986.

Platelet-Derived Growth Factor (PDGF)

It had been shown in the 1920s that animal serum contained factors favourable to the growth of cells in tissue culture. This is one of the awkward problems facing those who oppose any use of animals for medical research and advocate instead the use of tissue culture cells as a substitute. It is true that there are many applications for which tissue culture cells are more suitable than whole animals, but cells nearly always need various substances derived from animal tissues to enable them to grow, and the most usual requirement is for serum. So the animals are still needed even if the actual experiments are performed on the cells.

Serum is the part of the blood that remains liquid after clotting. It is slightly different from plasma, which is the part of the blood left when the cells are removed by centrifugation. Plasma can still clot, but it does not contain the blood platelets that release their content of secretory factors during clotting. It was shown as late as 1974 that the platelets were the source of the principal growth factor in serum, and it henceforth became known as *platelet-derived growth factor,* or PDGF. As PDGF is not nearly as abundant as NGF or EGF, its purification was a heroic task, requiring batches of platelets from several hundred litres of human blood as the starting material, and the use of numerous chromatographic columns. Complete purification resulted in an increase in specific activity by a factor of 10^7. This means that 1 part of PDGF had to be separated from 10,000,000 parts of other proteins. It was achieved simultaneously in 1979 in two labs, one Swedish and the other American. At that time, the determination of the amino acid sequences of proteins was still quite difficult, and it was not until 1983 that the sequence was obtained, again by two labs, one European and the other American. But this was a sequence that nobody would ever forget! Computer searching immediately showed that the primary structure of PDGF, the growth factor, was very similar to that of a gene called *v-sis;* and *v-sis* was one of the then newly discovered oncogenes: a gene that could cause cancer.

It has since turned out that several of the growth factor receptors, the signal transduction kinases, and the transcription factors that they regulate are oncogenes. In some cases they can cause excessive growth simply by overexpression; in other words, if there is too much of the protein present, the pathway to which they belong becomes overstimulated. Usually, however, the oncogenic character of the oncogene is due to a mutation that changes the biological activity of the protein. In most cases the cancer-causing mutation makes the protein active all the time, whether or not the growth factor is present. This means that the cells containing the mutation no longer need stimulation by the growth factor in order to divide. Because the signal transduction pathway is active all the time, they do not need the factor or the proximity of other cells producing it, and they will continue to grow without limit in any part of the body.

The Zoological Garden of Growth Factors

One of the characteristics of scientific research that is so hard to explain to politicians and administrators is that things never quite work out the way you expect. Large and expensive programmes of "targeted" research, designed to improve economic competitiveness or to cure particular diseases, often yield large masses of stodgy data with limited utility. Real novelty usually arises from small-scale work in some backwater previously thought hopelessly unfashionable and unfundable. This has certainly been true for the discovery of the growth factors. This section briefly introduces four of the families of factors that have been found to be of particular importance in the embryonic development of animals. We shall meet these various and confusingly named factors again in Chapters 5 and 8; the object of the present section is to show where they actually came from. They are the fibroblast growth factors (FGFs), the transforming growth factor betas (TGFβs), the bone morphogenetic proteins (BMPs), and the "Wnts" (pronounced "wints").

One thing that has become very clear in the AC era is that a large proportion of gene products in vertebrate animals come in families of related molecules. If there are 10 types of a particular factor, then there are 10 genes coding for them. These genes are similar but not identical; in other words, a proportion of the nucleotides in their sequences are the same, and a similar proportion of amino acids in the corresponding proteins are the same. This is true not just of growth factors but also of enzymes, structural proteins, and transcription factors. Because there can easily be 5 or 10 related genes for any particular class of factor, the names given to factors can be extremely variable and depend on the precise route of discovery. After a few members of a gene family have been discovered, there is usually a situation of extreme chaos and inconsistency in the nomenclature. So a group of the great and the good meet together and issue a portentous statement in a fashionable scientific journal such as *Cell* to regularise the situation. Usually this means that one family name is adopted and the genes are numbered in order of discovery.

As we saw in Chapter 1, the story of the FGFs started in 1939 with the discovery that brain extracts could promote the growth of tissue culture cells. The relevant reference is of some interest because it is sometimes misquoted as "Trowell, Chir, and Willmer, 1939." Willmer was the senior author, a well-known authority on tissue culture and developmental biology in Cambridge, England. Trowell was the first, or junior, author who had actually done the experiments. But who was Chir? Actually, he did not exist at all. In those days the *Journal of Experimental Biology* used to publish the qualifications of each author with their names. As Trowell had a surgical degree, his name was followed by "B. Chir" (i.e., Bachelor of Surgery), which was taken by some careless indexer to be a third author. Rarely can a citation profile have had such a slender basis!

The original observation was more or less ignored until the 1970s, when serious attempts to purify the growth-promoting activity began. These were unsuccessful because of the low abundance of the factors, and it was not until the mid-1980s that the protein purification technology became good enough to achieve success. The breakthrough in this case was the discovery that the factor would bind with high affinity to heparin, a substance already very familiar as an anticoagulant, a substance that will prevent blood from clotting. This meant that a good purification could be achieved using a chromatographic column consisting of immobilized heparin. FGF was purified from brain and pituitary gland, and the amino acid determined. At the time of writing, there are 15 different FGFs known. As suggested in Chapter 1, FGFs have turned out to be very important in embryonic development. They are needed not only for mesoderm induction but also for the formation of the posterior ("trunk-tail") part of the body, for tooth development, limb development, and the formation of blood vessels.

Tissue culture cells can sometimes be transformed into cells with cancerlike behaviour after infection with certain viruses. This is usually because the virus is carrying an oncogene that is a modified version of one of the normal host genes. Cells showing the cancerlike behaviour are called *transformed*. It was found in 1980 that the transformed cells could themselves produce a growth factor stimulating the growth of other, more normal, tissue culture cells. By a great stroke of luck, it was found that this factor stimulated the EGF receptor, which was the only growth factor receptor known at the time. Purification during the 1980s showed that there were actually two factors in the medium that acted synergistically with one another. They became known as transforming growth factor α and transforming growth factor β (TGFα and TGFβ). Further study showed that the two factors were biochemically and biologically quite different from each other. TGFα was indeed rather similar to EGF, but TGFβ had a totally different molecular structure and, on its own, proved to be inhibitory to growth for most cell types. By this time it had also become clear that neither of the TGFs was a special product from transformed cells but that they were both widespread products of normal tissues. So TGFβ is not "transforming," it is not a growth factor, and it is unrelated by any chemical criterion to TGFα! Nonetheless, its name has stuck and it is too late to change it now. In fact, TGFβ became the prototype member of a new and diverse "superfamily" of growth factors; prominent among these are the bone morphogenetic proteins (BMPs).

The BMPs were discovered by following up a remarkable property of *demineralized bone matrix*. Bone consists of cells called osteocytes surrounded by a considerable amount of extracellular matrix material heavily impregnated with calcium phosphate. Demineralized bone matrix is bone that has been treated with acid to remove the calcium phosphate, and it consists mainly of insoluble protein. In 1965 it was shown that implantation of demineralized bone matrix into various sites of various animals would provoke the formation of new bone by the neighbouring cells. This was all the more remarkable because the process appeared to recapitulate the normal process of bone formation in the embryo, which in most parts of the body involves the initial formation of cartilage and the subsequent replacement of the cartilage by bone. This sort of complex biological effect is a big problem for the protein biochemist because protein purification requires the ability to do a large number of assays, or measurements of biological activity, on numerous column fractions. The sort of assay that biochemists are used to is something like an enzyme assay in which the substrate is added to each tube and the rate of formation of product is measured by a colour change. This is quick, simple, and cheap. A typical growth factor assay is rather more complex, as it usually involves treating living cells with the fractions and measuring the rate of DNA synthesis by the incorporation of a radioactive DNA precursor. The BMP assay was an absolute nightmare because it involved stripping the demineralized matrix of activity by treating it with very aggressive substances like guanidine hydrochloride, adding back the fractions to samples of the matrix, implanting these under the skin of rats, and waiting several days to look for the development of cartilage in the implant.

The BMP assay resembles the sorts of things people tried to do when attempting to purify inducing factors from embryos. Not only is it very labour intensive and time consuming, but it is not quantitative, and this imposes severe problems on the biochemist who wants to be able to account for the fate of all of the units of activity of the protein during the purification. However, the BMPs were purified, or nearly purified, and the work was undertaken by Genetics Institute, Inc. (GI), one of the new biotech companies that sprang up in the 1970s and 1980s. A company has the resources to do things that are not possible for an individual academic supported by ordinary research grants. However expensive a protein purification turns out to be, it is nothing compared to the enormous costs of clinical trials or of getting a new drug through the regulatory hoops that are imposed by the U.S. Food and Drug Administration. So a large team at GI tackled the problem of BMP purification head on, using the cumbersome rat implantation assay. They did not get an entirely pure product, but they pushed the purification to the point at which they had a mixture of proteins that was not too complex, and then they carried out limited amino acid sequencing on many of the components. This enabled them to design oligonucleotide probes to isolate the BMP genes from suitable libraries. There turned out to be several BMPs, but most of them formed an obvious biochemical family that by structure and sequence clearly belonged to the TGFβ superfamily of factors.

The main interest of GI was to isolate factors that could be used to promote the healing of recalcitrant fractures of bone. Despite all the skills of the modern surgeon, this can still be a considerable practical problem. Most fractures heal, but some do not, for reasons that are often somewhat obscure. Surgeons frequently resort to grafts of bone from elsewhere in the patient, and it now seems clear that these work because they secrete BMPs that stimulate the cells in the vicinity to make more new bone. Surely there was money to be made from a substance that could be painted onto the fracture rather than resorting to the cost and inconvenience of a graft? Time will tell the ultimate utility of the BMPs in this regard, but they certainly do work to promote fracture healing both in animals and in human patients. They are also of considerable interest to the developmental biologist. As might be expected, BMPs are necessary for bone development, but some of them are also very important for other processes, including, as we shall see, the dorsal-to-ventral (back-to-belly) organisation of the entire animal or human body.

The BMP story is one of dogged persistence in working towards an objective that could be seen, at least some way, in advance. The story of the Wnts is one of remarkable coincidence. Among the viruses that can cause cancer in animals is mouse mammary tumour virus (MMTV). This is a DNA virus that can integrate into the genome of the host. Unlike some other tumour viruses, it does not carry an oncogene but works merely by virtue of activating a host gene at its insertion site. It can integrate into many different sites in the genome, and at most of these there is no effect. However, there were two sites where integration of the virus into the genome of cells of the mammary epithelium was found to cause formation of a mammary tumour. It was presumed that a host gene, not normally active in the mammary gland, was being activated at these particular sites and that this led to the cellular transformation. The genes became known as int-1 and int-2 for integration sites 1 and 2. After the inevitable long hard cloning exercise, the host genes were identified: int-2 turned out to be a member of the FGF family, later called FGF-3; int-1 was found to be closely related by primary sequence to a gene from the fruit fly Drosophila called wingless.

As we shall see in Chapter 8, the understanding of Drosophila development has led to the identification of many genes that also control the development of vertebrates. The wingless gene is a segment polarity gene, meaning that it is concerned with the formation of the repeating segmental pattern of the Drosophila larva and adult. The primary sequence of the factor made it clear that it was a growth factor of a novel type. In this case the nomenclature committee decided to credit both lines of discovery by calling the new factors Wnts, taking letters from both wingless and integration site. Subsequent work has resulted in the isolation of many other members of the Wnt family.

There are many other growth factors, equally important in their contexts, not discussed here. In particular, there is a large class of *interleukins,* which are factors produced by lymphocytes and acting, in the main, on other cells of the immune system. Then there are several *neurotrophic growth factors,* of which NGF was the prototype, and whose functions lie mainly within the nervous system. There are the *interferons,* which are produced by many cell types in response to viral infection, and there are the *haemopoietic growth factors,* or colony-stimulating factors, that control the differentiation of the cells of the blood. These factors are perhaps not quite so important for embryonic development as those mentioned above, but they are very important in the marketplace because they have properties that can be put to immediate clinical use.

Megabucks

In no area of academic molecular biology has the rush to commercialise basic knowledge been as frantic as in the area of growth factors. Hundreds of American postdocs, having slogged their guts out for several years cloning some novel factor, have decided that they would rather make money out of their skills than use them merely to get their lab head's name into *Cell,* the most prestigious of bioscience journals. They have set up new companies around a factor, or group of factors, and often have been able to attract substantial investment funds to develop them for some possible therapeutic purpose. In some cases companies have been set up by senior scientists jaded by years of grant applications and departmental administration. By 1993 there were in the United States 1,272 biotech companies employing 80,000 people directly with annual sales of $6 billion. Arthur Kornberg, a Nobel Prize winner and one of the founders of the DNAX company, has described biotechnology companies as "the new alchemists, with the ability to turn DNA into gold."

In fact, the gold mainly flows from investors into the companies, as only a few of the biotech startups of the 1980s have become really profitable. The basic case for investment is undeniable: molecular biology has provided a potential cornucopia of new molecules, with high biological activity, whose normal functions are to control the most intimate biological functions of cells. There must eventually be enormous payoffs from the applications of all this new knowledge in medicine, veterinary medicine, and agriculture. The presentational skills of the scientists have attracted substantial investment, but the fact is that this is an expensive technology to develop and the commercialisation process has begun very early, when there are still many gene products to be identified and much basic information to be collected.

The most successful company has probably been Amgen, set up in 1980, which was early able to market two novel products of high efficacy. Erythropoietin is a haemopoietic growth factor that stimulates the formation of red blood cells and is a uniquely effective treatment for anaemia, particularly for those on haemodialysis. Granulocyte colony stimulating factor (G-CSF) is another haemopoietic growth factor, which stimulates the growth and differentiation of white blood cells and is useful in the treatment of various blood disorders. Notable among these today is the bone marrow failure brought about by medical oncologists when they treat their patients with anticancer drugs. Such drugs are often very toxic to the blood-forming cells, and so an agent that specifically promotes recovery of the marrow is very useful. The initial investors in Amgen who put in $9,000 each were able to recover $130,000,000 10 years later. There have been many other successes. For example, the interferons, marketed by Biogen and Genentech, are substances useful in antiviral therapy and in one particular form of leukaemia. Interleukin 2 (Chiron) is a growth factor for lymphocytes useful in treating renal cancer. Ceredase (Genzyme) is a uniquely effective treatment for Gaucher's disease, a rare inherited condition in which an enzyme is lacking that is needed to break down a particular type of lipid. None of these products could have been made without molecular cloning, and in fact, some were not even discovered until the technology made it possible.

Although a number of biotech companies have proved reasonably profitable after the first few years, the majority have never hit the big time in terms of actually producing and marketing a product. However, there are other ways to turn DNA into gold than this traditional route to business success. It often seems as though the main aim of a new biotech company is to be bought up by one of the big pharmaceutical houses at some point before its start-up funds run out. This will make a huge profit for the initial investors and guarantee a future for the staff, the labs, and probably the research programme. The principal asset of a small company is usually its own scientific staff. At first sight this may seem surprising. Why should anyone bother to buy the whole company when they could just hire the individuals directly? But the answer lies in the traditional reluctance of academic scientists to enter industry. The junior lab members in a university, who are all on fixed-term contracts, always want to be academics themselves and have a big lab and loads of grant money, like their supervisors. Because there are never enough slots to make this possible, many have to settle for something less. This may mean going to a "teaching university" (the true meaning of this apparent tautology will become clear in Chapter 9) or going into industry. But the old-style pharmaceutical industry is seen as very unattractive, involving loss of freedom and uninteresting work.

This is perhaps somewhat unfair, because an academic molecular biologist spends all day pipetting small volumes of colourless liquids from one tube to another, and his industrial counterpart does exactly the same; but it is nonetheless the popular perception. The new-style biotech companies are rather more attractive because they often have close links with universities, having been set up by academic scientists, and they are often run in quite an open way, with little restriction on publication and the opportunity for senior staff to recruit postdoctoral fellows in the same way as they would in a university. All this means that they generally find it easier to recruit able staff, and these staff are a real asset when the time comes to sell out.

The other principal asset the companies have to sell is their patents. Patents have always been very important in the pharmaceutical industry. If a small company has cloned a particular gene that may be of interest to a large company, and it has taken out a clutch of patents for every conceivable derivative and application of the gene product, then it can save a lot of lawyers' fees for the large company just to buy the whole lot as a package. Finally, there is the remarkable fact that the scientific advisory board of a company can have a real value to its buyer. It is traditional for biotech start-ups to persuade some well-known senior academic scientists to sit on a scientific advisory board. Sometimes the advisors have a real role in assessing the progress and direction of the work, but usually they are just there so that the company can put the names on its notepaper. The lure of having these great names on the notepaper of the purchasing company can be strong. The people concerned may be Nobel Prize winners, or at least regular publishers in the journal *Cell*. They are the superstars of the life sciences, and the sale of their names can command a substantial price. Of course, as there is nothing to stop a superstar from being on the scientific advisory board of several companies, they can even be sold many times over!

Patents

The object of the patent system is to provide some incentive for innovation. In a market-dominated economy, there is no incentive for company A to devise a new product in a situation in which the development costs far exceed the manufacturing costs. The reason is that as soon as company A had developed a product, companies B, C, and D would move in and manufacture it just as cheaply, but without having had to pay out all the money required for the research and development. They can then undercut company A, with the result that company A will go bust and companies B, C, and D will make lots of money. Once the first product became obsolete, they would look round for another innovator to copy, but of course, there would be no more innovators in such a system because the shareholders of companies like A would by then have insisted that they behave instead like B, C, and D.

Discoveries may be protected to a limited degree by secrecy, but in pharmaceuticals it is fairly easy to establish the nature of a product by chemical analysis. A method of synthesis may be more difficult to copy, but even hiring some good chemists to do this is vastly cheaper than going through the expensive hoops of clinical trials and regulatory approval. The patent system is a long-established way whereby the state interferes with the free market in the public interest. It grants a monopoly position for a particular product to its inventor for a certain period of time. During this time nobody else may manufacture the product unless they buy a licence from the inventor. While a drug is under patent protection, it can be sold for whatever price the market will bear. A unique and effective product can command a truly impressive price. Agents such as the interferons and haemopoietic growth factors can be sold for hundreds of dollars per dose, and of course, people (or their insurers) will pay such prices for a product if it is really effective and means the difference between life and death.

Particularly in the field of pharmaceuticals, many members of the public cannot understand the justification for this system. But the fact is that almost everything one does in research is a failure. Companies have to examine thousands of substances for every one that makes it to the market, and although the early stages of research are quite cheap (by commercial standards), the stages of advanced animal testing, clinical trials, and the many steps needed to gain regulatory approval from the U.S. Food and Drug Administration are extremely expensive. It currently takes about 10 years and costs about $300 million to bring a new drug to the market. So a patent system is certainly necessary to provide some incentive for innovation. Patents themselves may be traded and can represent the large part of the value of a biotechnology company.

The main problem about patents is what sort of claim constitutes a reasonable ground for protection. This is a thorny matter, and there are large numbers of patent lawyers who make a good living sorting such things out. In principle a patent can only be granted for something that is novel, useful, and not obvious to everyone. There have been great swings in the threshold for patentability over the years, but in recent times it has been felt that it is generally easier to get a patent in the United States than in Europe. The biotechnology business has spawned large numbers of patents and many acrimonious and expensive legal battles over them. An early success was the patenting of the molecular cloning technology itself. This was done by two of the innovators, Stanley Cohen, from Stanford University (a different individual from the Stanley Cohen of EGF fame), and Herbert Boyer, from the University of California, San Francisco. Academic researchers do not need to worry about this, but companies using the technology to make products for sale have to pay an annual licence fee of $10,000. This is quite a small fee, being less than the consumable costs of keeping one molecular biologist at the bench for a year. This means that it is not worth contesting legally, so the companies continue to pay up. It was well worth it. Over the lifetime of the patent, which was between 1974 and 1997, the two universities netted about $170 million.

One of the first profitable biotech products was erythropoietin, the haemopoietic growth factor that stimulates the formation of red blood cells. Amgen had a patent for the purification and manufacture of recombinant erythropoietin. But another company, Genetics Institute, had an earlier one for the purification of erythropoietin from urine, plus a conception of cloning the gene. The legal battle went on for four years. In the end Amgen won and cornered the market for erythropoietin, something that has contributed substantially to its profitability. Another disputed patent was that for the polymerase chain reaction. This is a method for amplifying stretches of DNA in the test tube, and it has multifarious applications. A patent for the method was granted in 1985 to the Cetus company. It was challenged in 1991 by Du Pont on the grounds that all of the steps were well-known enzymic reactions and the procedure had been envisaged some years earlier. The award of the 1993 Nobel Prize for Chemistry to Kary Mullis of Cetus probably helped Cetus's case along, and they eventually won. However, Cetus had already been bought by another biotech company, Chiron, who then sold the PCR patents to Roche. All this wrangling over patents is enormously profitable for the patent lawyers, and because a lot of novel and complex scientific issues are involved in the court cases, it must by now represent a highly attractive career switch for a scientist with a taste for advocacy.

Whether the patent system should be extended to naturally occurring substances like growth factors, or to the sequences of genes that code for them, has been a subject of hot debate. Historically, natural substances could not be patented, although new methods of synthesis, extraction, or manufacture could be. Recently, attempts have been made to patent large numbers of partial or complete gene sequences from the human genome. In many cases nothing is known about the protein product, and in most cases all that can possibly be known is an inference about its biochemical function based on the similarity of the primary sequence to that of other known genes. For example, an unknown sequence might be identified as a kinase (an enzyme transferring a phosphate group) or a receptor (cell surface molecule binding a growth factor, hormone, or neurotransmitter).

But even this tells us nothing about what its role is in the organism or whether any useful products might be made from it. For these reasons, it is surprising that attempts to patent human genes have got as far as they have.

In fact, the mad rush for intellectual property rights has now reached the most absurd extremes. In the United States today, companies are even buying up mutations that look as though they might be in interesting genes, in the hope that this will give them some exclusivity over the cloning and sequencing of the gene. There exists a fashionable organism called the zebra fish, which is quite favourable for doing genetic experiments, and a large number of mutations affecting embryonic development have recently been isolated. Because the mutations were induced by treatment with chemical mutagens, nobody knows in which gene a particular mutation lies, whether it is a known or unknown gene, or even whether the mutation lies within an actual gene or in a noncoding part of the DNA responsible for regulation of gene activity. The process of actually cloning a gene starting from a chemically induced mutation is a long and difficult task. But in the scramble for "proprietary positions," even the mutations of the humble zebra fish can apparently be regarded as tradable assets.

Despite the considerable medical and financial success of some products like the interferons and the haemopoietic growth factors, it is likely that unmodified gene products are never going to be that important as pharmaceuticals. Some protein hormones have been used therapeutically for a long time, the best known being insulin, self-injected every day for decades by many diabetics. In the past some protein hormones were extracted from animal tissues, and more recently they have been made by genetic engineering methods. But the main problem with any bioactive protein is that it has to be administered by injection. Proteins cannot normally be taken by mouth because the acid of the stomach and the digestive enzymes of the intestine will destroy them. They will be absorbed as amino acids or small peptides rather than as whole protein molecules, so all the specificity and bioactivity will have been lost. For growth factors there is the additional problem that they are often rather short-lived in the blood circulation because they are substances designed to have a local rather than a systemic action. For gene products such as receptors, enzymes, or transcription factors, their normal location is inside cells; they cannot be used as therapeutic agents unless some way is found to introduce them into the cells of a living person. This is the much vaunted "gene therapy" in which a gene may be introduced into cells using a virus. But it is only likely to be possible to introduce genes into a few cells rather than into every single cell of a particular type. So the types of disease that could be treated will be limited to those in which the restoration of normal function in a minority of cells is likely to be effective. This would be the case, for example, with a metabolic defect in which one enzyme is missing and the substance it normally makes can diffuse freely from cell to cell. Here, the introduction of the missing gene into a few cells might work because those cells could feed their neighbours with the missing substance. For conditions like cancer, in which every single diseased cell needs to be eliminated, the hype surrounding gene therapy seems likely to exceed the effectiveness for a very long time to come.

The pharmaceutical industry has had most of its successes with relatively simple organic chemical substances that can be taken by mouth, and this will probably continue. The importance of molecular biology and genetic engineering is not so much that it will directly create a lot of new pharmaceuticals, but that it will identify all the gene products and tell us something about them. This means that there are already thousands more potential therapeutic targets for intervention than there used to be. The large pharmaceutical companies are all worried about running out of steam and running out of new products, but molecular biology promises a route to discovering products that can stimulate or inhibit any one of the 10^5 gene products in the human body. This is probably why they are willing to pay good money for biotech companies. Ultimately, they do not necessarily want the companies' flagship proteins, or even their patents or headed notepaper. They want to buy into molecular biology itself, because only that can secure their future into the next millennium.

Postscript: Molecular Biology, BC and AC _____

(*Note:* This part is for the benefit of readers not familiar with the basic ideas of molecular biology. It is perhaps less interesting than the remainder of the book, so if you do not need to read it, please skip.)

The basic facts of molecular biology are as follows: genes are made of DNA (deoxyribonucleic acid), and they are to be found in chromosomes in the cell nucleus. DNA consists of long threadlike molecules made of building blocks called *nucleotides*. There are four types of nucleotide (called A, T, C, and G), and they can occur in any order. The molecular structure of DNA consists of a double helix in which the sequences of nucleotides on the two strands are complementary to each other. This is because of the pairing rules stating that A on one strand must pair with T on the other, and C on one strand must pair with G on the other. So, although the sequences of nucleotides on the two strands are different, the pairing rules ensure that the sequence of one strand determines the sequence of the other. One strand of the DNA encodes the gene, and the other strand, sometimes called the *antigene strand,* is necessary as a template for replication. The DNA must replicate itself before each occasion that the cell divides in order to make a perfect copy of the genes for each daughter cell.

The sequence of nucleotides in a gene codes for a particular protein molecule that is the product of that gene. Proteins do all the important jobs in the cell. They make up its structure, they regulate substances that enter and leave, they catalyse numerous chemical reactions, and they control the activity of the genes themselves. Proteins are big molecules consisting of chains of amino acids of which there are 20 different kinds. The "coding rule" for converting DNA to protein is that each triplet of three nucleotides codes for one of the 20 amino acids. There are 64 possible triplets, of which 61 code for amino acids and 3 are "stop" signals. Since 61 is more than 20, most of the amino acids can be coded for by more than one nucleotide triplet. The identity of the nucleotide triplets that code for each amino acid is known as the *genetic code* and was established in the early 1960s. When a gene is active, molecules of RNA called messenger RNA, or mRNA, are synthesized complementary to the gene strand of the DNA. RNA (ribonucleic acid) is similar to DNA except that it is normally single-stranded and it contains a base called U in place of the T found in DNA. During protein synthesis the sequence of nucleotides in the mRNA is used to assemble the corresponding sequence of amino acids and to join them together to form a protein. This complex task is carried out by intracellular particles called ribosomes. Most of the activities of a cell such as its metabolism, response to stimuli, or structural maintenance are carried out by proteins. So the nature of a cell—for example, whether it is a nerve or a muscle cell—depends on the proteins from which it was assembled and therefore on the particular subset of its genes that are active in its nucleus.

All these facts were known, or guessed, almost as soon as the structure of DNA was established in 1953. However, the real key to modern genetic engineering lies not in the theoretical knowledge but in the practical techniques of molecular cloning, otherwise known as recombinant DNA technology, that were invented in the early 1970s. In fact, we can date the beginning of the cloning era precisely from a paper published in 1973 from the lab of Herbert Boyer at the University of California, San Francisco (UCSF) entitled "Construction of Biologically Functional Bacterial Plasmids *In Vitro.*" It described a set of techniques for putting a piece of DNA into bacteria and growing it up so that a useful quantity of a single gene could be isolated in a test tube. If 1973 is designated as year 0 of the molecular biology calendar, then the structure of DNA was discovered in 20 BC (before cloning) and at the time of writing, we are in the year 25 AC (after cloning). The first quarter century of the AC era has been one of explosive growth and success for laboratory-based biology, and in most cases, the success has derived from plugging the techniques of molecular biology into a preexisting biological problem. The result has been an understanding of the biological phenomenon not just at the level of whole organisms and cells but also at the level of genes and proteins. This has happened for such fields as diverse as virology, neurobiology, cancer, immunology, and developmental biology. In each case understanding the molecular basis of the events makes it possible to think about new types of practical application that are also based on the new technology of genetic engineering.

The importance of molecular cloning lies in the fantastically tiny quantities of individual genes and gene products that exist in nature. Each organism has a *genome,* or set of genes encoded in its DNA. The DNA is found in the nuclei of all cells in the body, and with a few exceptions, the genome is the same in every cell. The human genome contains about 100,000 genes, each of which consists of one or more stretches of DNA coding for a particular protein and which are separated by large amounts of noncoding DNA. Like most animals and plants, humans are diploid, meaning that they get half their chromosomes from the mother and half from the father. So each cell contains just two copies of each gene. A million cells contain 2 million copies of a particular gene. This may sound like a lot, but it is very little in terms of chemistry. This is because a single gene only weighs about 10^{-18} grams. The smallest amount of DNA possible to see with the naked eye is a few micrograms, where 1 microgram = 10^{-6} grams. To make enough of a single gene to be able to just see it requires at least 100,000 tissue culture flasks to grow enough cells to use as the starting material. This meant that in the BC era the only biochemical studies possible on genes were on those that occurred naturally at high copy numbers. There are not many of these, but they include the genes coding for the structural RNA of the ribosomes, which formed the not very interesting topic of my own Ph.D. thesis.

Molecular cloning overcomes this problem and allows any gene to be obtained in useful chemical quantities. It takes advantage of the inherent biological property of DNA that individual molecules can replicate and form many precise copies of themselves. It is in fact now possible to multiply up molecules of DNA literally in the test tube, using a method called the polymerase chain reaction (PCR). But this can only handle quite short sequences and is somewhat prone to error, so the methods of cloning used now are still essentially very similar to those originated in 1973. This means that the sequence of DNA encoding the gene of interest is chemically joined, or spliced, to a *vector,* which is a molecule of DNA that has the ability to replicate itself in a bacterial host. The vector is usually either a plasmid, which are small DNA circles that exist naturally in bacteria, or a phage, which is a type of virus capable of infecting bacteria. The tools for cutting and splicing DNA molecules are enzymes that were discovered in the 1950s and 1960s by the small group of pioneer molecular biologists who worked out the mechanisms of DNA metabolism. The three essential types of enzyme are the *polymerases,* the *restriction enzymes,* and the *ligases.* Polymerases synthesize new strands of nucleic acid complementary to existing ones. Restriction enzymes cut DNA at particular specific sequences. Ligases join one piece of DNA to another.

The terms *recombinant DNA* or *genetic engineering* refer to the ability to assemble novel DNA molecules at will using these various techniques.

Together with the enzymes and the cloning vectors, there are two other essential ingredients of the molecular cloning revolution: sequencing and hybridization. Modern rapid DNA sequencing methods enable workers to determine the sequence of nucleotides in any piece of DNA. The techniques of nucleic acid hybridization allow the identification of minute quantities of specific DNAs or RNAs containing a nucleotide sequence complementarity to that of a radioactive or chemically tagged "probe" molecule. Both these methods were developed at about the same time as molecular cloning, and together, the whole box of tricks makes up most of modern molecular biology.

If you are interested in any gene or its product, molecular cloning is now an essential first step in your work. Once you have cloned your gene, you have a test tube containing milligram quantities of it. You can then find its sequence and do computer searches to find what biochemical class of protein it codes for and what other molecules it is related to. The clone also enables you to make probes. These are bits of nucleic acid (DNA or RNA) synthesized from the noncoding strand of the clone that can be used to identify the natural messenger RNA, measure it, or localise it in cells, tissues, or embryos, with methods depending on nucleic acid hybridization. The level of a particular mRNA is a direct measure of gene expression, meaning the activity of the gene in a particular cell type or a particular developmental stage, or following particular treatments.

Perhaps most important, once you have a clone, you can make limitless amounts of the protein product by introducing your clone into bacteria or tissue culture cells. Before molecular cloning, the only proteins that could easily be obtained in pure form were those that naturally occurred in high concentrations. Most proteins, and almost all of the interesting ones with high biological activities, exist at very low levels of just a few thousand molecules per cell and require heroic quantities of starting material for purification. But unlike the normal gene, a clone you have introduced into bacteria or tissue culture cells will be present at many copies per cell and will contain very powerful regulatory sequences around the gene to ensure a high level of protein synthesis. So the recombinant protein will be made as a significant fraction of total protein, which makes it much easier to purify. With the pure protein, you can make antibodies by immunizing mice or rabbits. Antibodies are essential tools in molecular biology as they can bind to any target molecule with high affinity and high specificity. This property can be used to measure the levels of the protein accurately, to localise it in tissues, cells, or embryos, or, if the antibody is a neutralizing antibody, to inhibit the protein specifically in biological experiments.

For all these reasons, the cloning of a gene is an essential first step to almost any research project in modern laboratory biology. It is also the source of much hard work and anguish. Genes are cloned by screening a *library,* which is a complex mixture of clones made from a primary source such as genomic DNA or the total mRNA from a particular tissue. Screening is usually done using a highly radioactive nucleic acid probe. This will only react with clones of complementary sequence and so serves to identify one particular type of clone out of a whole library. The probe is prepared either corresponding to the amino acid sequence of an interesting protein or to the sequence of related genes of interest. Sometimes genes are cloned by some screening method based on the biological activity of the protein, which is exciting because the clones that come out may have novel sequences that are completely unknown in advance. However, it is also more difficult and risky than screening with a probe. Because cloning a gene is such a decisive step, there are frequently races between labs trying to do it first. It has also increased the frustration level of academic science considerably because nobody now wants to read merely about a gene being cloned. So the time and effort spent isolating the gene and making the specialised types of derivative clone required for a range of experiments represent a huge amount of work, expense, and risk. And all this has to be undertaken before the real experiments can even begin.

CHAPTER

On the Catwalk:
Publication and Presentation

The Journal

In some ways scientists are not at all like actors. Their personalities can often be quite introverted and their level of social skills somewhat below the national average. In other respects there are more similarities, the term *prima donna* being not uncommonly applied to the more awkward research stars in an institution. Perhaps the closest point of contact is in the craving for publicity. Not publicity in the national media, which are often shunned and regarded as irresponsible, but publicity in the specialised media of science.

The results of research are reported in scientific journals. A typical journal covers a specialised branch of research—for example, developmental biology—and appears perhaps once a month. It consists of a number of "papers" reporting the results of original experimental work. These papers are written in the most extraordinary stilted and standardised manner making them completely opaque to nonspecialists in the subject. Everything is in the passive voice and past tense, and the arrangement of the paper always conforms to a precise prespecified format. The opacity is not maintained with the deliberate intention of excluding the general reader; it is rather that scientists feel they have to write in this straitjacket because everyone else does. Journal editors tend to be more liberal than their authors and sometimes make vain attempts to encourage authors to use the active voice or adopt a more imaginative presentation, but so far without much success.

A scientific paper usually consists of the following components:

1. *A Title.* This is often long and cumbersome, as it is intended to communicate the substance and mention as many of the key terms as possible.

2. *The Author List.* Long gone are the days when senior scientists routinely performed their own experiments, so multiauthor papers are the rule. The general rule is that the first author is the one who did most of the experimental work, the last author is the lab head who had the idea and obtained the money, and the remaining authors have contributed particular specialised parts to the experimental work. Three to six authors is quite normal, and an author list can easily run to 20 people for complex pieces of work.

3. *An Abstract.* This is a 200- to 300-word summary of the results. Abstracts are published on various databases such as Index Medicus or the Science Citation Index, which are available on every networked computer in a research institution. Quite often when checking the contents of a publication, the abstract is all you need.

4. *The Introduction.* This states the problem, briefly refers to recent previous work bearing on it, and says what the new experiments are. It is important in this section to cite the previous work of anyone who might review the manuscript before acceptance.

5. *The Methods.* This is a detailed account of the actual experimental techniques. The level of detail is very high because readers skilled in the art should be able to use this information to repeat the work in their own labs. So, for example, the sources of complex substances such as antibodies would all be listed in case there are differences between the products of different manufacturers.

6. *The Results.* This is the real meat. The results of the experiments are described in detail with tables of figures and graphs to summarise numerical data and photographs to show "typical" (i.e., the best) specimens.

7. *The Discussion.* The main conclusions are stated and the results are put in the context of previous work. The reasons for any discrepancies are discussed, and the likely future course of the work is briefly indicated.

8. *The Acknowledgements.* This says who paid for the work and usually thanks people who provided clones or antibodies or who read the manuscript before submission.

9. *The References.* This lists all the previous work cited in the article and usually consists of 20 to 50 references to other journal articles. They are in a standardised form appropriate to the journal, usually giving the authors, title, journal, year, and page numbers.

Scientific journals are not all equal. There is a steep pecking order from the anxiously perused "fashion journals" at the top to rarely consulted non-English language publications at the bottom. The raw material for deciding on the importance of a journal is to be found in the reference lists of its papers. If papers in a journal are cited regularly by other journals, then it is generally thought that it must be pretty good. This is made quantitative in the form of the so-called *impact factor,* which is the number of citations to all articles published by the journal in one year, divided by the total number of articles published in that year. The published impact factors are averaged over the past two years, so journals lose out if the readers only wake up to the significance of the contents by year three. Impact factors vary from about 30 to 40 for the fashion journals to around 1 to 4 for the less good, but still just about respectable ones. Really low class journals are not even included in the calculation by the Institute for Scientific Information, which is the body that publishes the Science Citation Index and all the data about impact factors.

Scientific literature does not just consist of original research papers. The volume of these is so great that there is a huge demand for short summaries and review articles. So there are several journals, notably the *Trends* series published by Elsevier, that are entirely devoted to short reviews of the current primary literature. Although one can obtain credit for writing reviews, it tends to be less than for research reports, and if you write too many, you are suspected as having lost your touch in real research. Scientists do not often write books. It is possible to advance your career with a good book, particularly one written for a general audience, but scientists do not want to read research reports in the form of books because they do not receive the same critical scrutiny and quality control as journal articles.

In the struggle for both personal and institutional advancement, a commanding position is held by the fashion journals *Nature, Science,* and *Cell.* This is because all scientists are constantly having to be assessed, and those who do the assessing are very busy, having a million and one other things to do each day in addition to reading long CVs attached to applications for grants, promotion, or new appointments. Young scientists think that the *number* of their publications is very important. And so it is. But busy senior scientists reading their hundredth CV can't be bothered to count to more than about 10; so if the list is longer than that, they just run their eye down it looking not at the content of the articles, and not even at their titles, but just at the *names of the journals* in which they are published. Preeminent are the fashion journals. After these rank the solid specialist journals. In my field they would be *Development, Developmental Biology,* and *Genes and Development,* followed by less worthy but still respectable *Mechanisms of Development, Roux's Archives of Developmental Biology, Development Growth and Differentiation,* and so on. Within each field, the workers have a detailed knowledge of the pecking order and know the impact factors of all the journals. Of course, nobody outside the field would have the slightest idea whether *Developmental Biology* or *Developmental Genetics* was more prestigious, and this is where the fashion journals score, because papers in *Nature, Science,* and *Cell* are universally recognised and accepted. They are in the currency of science what the deutsche mark is in the world of real money. *Nature* and *Science* are perhaps faintly familiar to the general public, as science journalists often consult them for news. Among biological scientists, *Cell* is the most fashionable of all, but for some reason it remains unknown to the general public. Indeed, one of the reviewers of this book recounted that when he excitedly announced to his family that he had published a paper in *Cell,* his grandfather thought it was *Sell!* and that he had changed his career to advertising.

Young scientists crave publication in the fashion journals with the same desperate intensity that their elders crave senior appointments or prestigious awards. They feel that it would really add significance to their miserable lives and that they would really be happy if only they could achieve this Nirvana just once (or just once more, since it is an addictive habit). They also believe what the editorial staff of the fashion journals themselves would like to believe, that the articles they carry really are much more important and significant than those in other journals. Of course, the actual truth is more prosaic. The fashion journals do publish some really important pieces of work that can be easily identified as such, and their high impact factors arise because these articles attract hundreds of citations per year. But the distribution of impact factors among the papers published by any journal is highly skewed, and the majority of articles will score well below the average value. In other words, most of what the fashion journals contain is actually the same sort of thing as the specialist journals carry but dolled up to look a bit special. Before being accused of sour grapes, I should perhaps put on record my own score so far of six papers in *Nature* and three in *Cell.* With the exception of the publication of the work described in Chapter 1, I do not think that any of the other eight publications was any better than work published in specialised journals at the same time; indeed, the citation performances bear out this belief.

Publications in the fashion journals not only identify "good people" in science, they also identify "good places." You can easily identify a good place because it contains good people, and these are the ones who publish their articles in "good journals." Even if you were not abreast of the latest impact statistics, you would have no doubt about which are the good journals because good people will keep telling you every day which are the good journals. If you are working in a good place, it is generally thought to be certain death to move to a place that is less good, or not at all good. This is because the panels awarding research grants or the editors of good journals will assume that you yourself are not much good if you have been obliged to move out from the handful of recognised good places around the world. This system does tend to make top-level academic biology somewhat introverted. Fortunately, every so often someone from a less good place does discover something important, or someone moves to a not so good place and makes it better, so while the system is certainly snobbish and self-contained, it is not entirely hermetically sealed.

Ordinary citizens might be surprised to learn that scientists are not paid for the articles they write. Sometimes they actually have to pay to have them published, particularly in journals published by commercial publishers rather than by learned societies. Even if they do not pay for publication, they are likely to have to pay for inclusion of colour plates or for reprints. Reprints are printed copies of the article that can be sent out to other scientists who write and request them, and although their use has declined somewhat since the advent of the photocopier, journals still make a lot of money by selling them to the authors. As a general rule, journals published by learned societies are better in that they do not usually require page charges and they often supply reprints gratis. On the other hand, their subscription price tends to be higher, so the scientists have to pay indirectly by allowing their library a larger slice of the institutional cake.

So how do you get your article into a fashion journal? This is no easy matter. Because publication is seen as so desirable, these journals receive 10 or 20 times as many manuscripts as they have space to include, and the submissions have to be culled by a savage process of reviewing (discussed below). Since it is so hard to decide which of 20 good papers is really more important than all the others, the manuscript must be not only good science but also *good journalism*. Some may detect a slight whiff of sour grapes here, but my claim for this proposition is based not on sentiment but on the fact that it can be experimentally tested. Indeed, it has been tested, by simultaneously submitting two papers to *Nature:* one being a piece of reasonably interesting embryology of the type that is quite often published, the other a piece of dilettante-ish nonsense, but written with sufficient panache to appeal to journalists. No prizes for guessing which one was published!

The story started at a postgraduate seminar hosted by my colleague Phil Ingham. The seminar was the last of a cycle of 36, so appropriately enough it dealt with evolutionary homologies of developmental mechanisms, a topic otherwise known as "life, the universe, and everything." There had recently been discovered some genes that are expressed in the head during early development both of the fruit fly *Drosophila* and of the mouse. This was very interesting because it suggested, contrary to all appearances and most previous speculation, that the heads of these two wildly different animals really were formed by some sort of similar mechanism. The discussion rambled on for a while, as it does when people discuss evolution, then everyone went home. A few days later I was explaining to a senior colleague, Chris Graham, about the very interesting embryological paper I was sending to *Nature*. He looked doubtful. I then summarised the content of our postgraduate seminar and pointed out that it might be significant that the newly discovered genes in question, together with the well-known Hox genes (already known to control head-to-tail patterning, and discussed below in Chapter 8), were maximally active at the so-called *phylotypic stage* of the two organisms. This is the stage at which members of each of the two groups, vertebrates and insects, are most similar to the other members of their own group. The similarity of gene expression patterns suggested that there was a *super*phylotypic stage that covered both groups, indeed covered the whole animal kingdom. "You know," he said, "You would be better off submitting a paper about that to *Nature*." Then he had a masterly idea. "You could call it the *zootypic* stage." We said simultaneously, "We could call the gene expression patterns the *zootype*."

I dashed off a short letter to *Nature* written in the most ridiculous florid language, claiming that the zootype was the fundamental character that distinguished animals from plants, and including quotations from von Baer, Geoffroy St. Hilaire, and other masters of nineteenth-century zoology. I knew that I wasn't important enough to submit it single-handed, and to have credibility there would have to be several authors. Chris Graham was an obvious choice, since it had been his idea. I then approached Phil Ingham, but he was already putting the final touches to his own short review article for *Nature* about the head genes, so declined, unwisely as it turned out. Then I asked Peter Holland, an up-and-coming young molecular evolutionary biologist. He was also on the head gene review but wasn't worried about the prospect of appearing in *Nature* twice and agreed. So now I had a fairly respectable slate of three authors, and that was enough to ensure that *Nature* would at least have the letter refereed rather than throw it straight in the bin.

Not only the fashion journals but all serious scientific journals have the papers sent to them assessed by experts in the field. In America this is called *reviewing*. Here in Britain, where we love sport so much, it tends to be known as *refereeing*. Refereeing is quite a significant chore, as it is done only by senior scientists and involves maybe 50 to 100 manuscripts each year. Usually, referees are not paid and are even expected to pay for the postage to return the manuscript. Referees' reports are anonymous, which enables them to be quite rude about papers they do not like. For a specialist journal, the referee needs to examine the paper and judge whether the experiments have been correctly carried out, whether the results are sound, and whether the work falls within the scope of the journal to which it has been submitted. For a fashion journal, there is a more important and more difficult judgement to make: Is the work of such galactic significance as to be eligible to grace the pages of *this* magnificent journal? In most cases the decision is necessarily highly subjective, and because of this, both the judgements and final decisions give rise to much bad feeling. Although the referees' reports are anonymous, it is often possible to guess the identity of the referee. One infallible clue is that if he complains about insufficient citations to the magnificent prior work of Bloggs, then Bloggs is your man. Usually, authors are careful to cover themselves by making sure that the manuscript includes at least one citation to anyone who is likely to be asked to referee it. Apart from this, when writing articles for fashion journals, they know they must stress the general and universal significance of the work. For example, a minute morphological study of the big toe of an earwig would have a title something like "Why Do Earwigs Walk Backwards?" and start off: "Since the time of Cuvier, the earwig toe has represented a fundamental anomaly." After rejection the paper will normally be resubmitted to a specialist journal, and all this nonsense has to be removed because it sounds pompous and will offend the next lot of referees.

Around this time my worthy embryological paper was rejected by *Nature,* so we toned it down a bit and sent it to *Developmental Biology,* in which it was eventually published. The zootype paper took ages being refereed, and when the two reports came back, they were pretty bad. "Lightweight," "obvious," "fanciful," and so on, went the referees. Normally, that would be the end of the story, but when I reread the editor's letter, I realised that we did after all have a chance because, even if the referees didn't like it, she did. This was not said in so many words, but an oblique reference to possible resubmission under very restrictive and special circumstances was easy to decode by those familiar with the subliminal analysis of editor's rejection letters. Things improved still further when the third referee's report came back. It was not entirely dismissive and said that the article might possibly be publishable after extensive revision. We were almost in! Chris, Peter, and I held several meetings going over the contents in great detail, covering ourselves against accusations of circular reasoning, collecting more supporting data from other people, and trying to make predictions that would establish the global significance of our argument. When we resubmitted, it was accepted!

It was to be published in a section of the journal called "Commentary." This is lower status than the original research reports, but I did not mind because there is only one "Commentary" article at the beginning of the journal, it has a prominent position, and since it was a bona fide refereed paper (rather than an invited review or an ordinary letter), I could still put it on my CV. What I had not appreciated were the particular advantages of the "Commentary" slot. For some reason, the article must be a whole number of pages in length. Normally, editors are savage in their demands for cuts, but on this occasion, because the editor obviously liked the piece, I was asked to *lengthen* it to fit it into three pages rather than two. The next surprise was that she said, "You know that you can have free colour for your figures in this section." Free colour! *Nature* and most other journals usually charge authors for reproducing colour plates. These charges are very heavy (around $1,000 per page) and are intended to be a deterrent, although it is remarkable how many authors seem able to find the money to publish endless colour plates. So we produced some extra figures, including one in colour, and sent them off. Unlike the main section, for which manuscripts must wait three to four months before publication, "Commentary" articles are published immediately.

The next surprise was to be rung up by the publicity manager for the Serious Disease Society and warned about the press release. *Nature* had fallen for the bait in a big way and was presenting our article as some sort of breakthrough. "What is an animal?—Dr. Slack and his colleagues have devised a new test for deciding whether something is an animal or not. . . ." Sure enough, I was then rung up and asked to do two radio interviews on the work. The SDS is smart enough to train its senior staff in interview technique, so I wasn't worried about this. I knew that the essence of an interview is to decide what you want to say, cast it into a few memorable sound bytes, and then say it on the air, regardless of the questions you are actually asked. This worked fine, so for both interviews, I was able to emphasise the huge advances that had recently occurred in developmental biology and explain how this would rapidly lead to better treatments for serious diseases, while mentioning the SDS's name several times. A few weeks later, some tiresome scientists wrote in to *Nature* saying how unoriginal our work was and how their ideas about what an animal is were much better. Such letters are often published in a correspondence column along with some comments from the authors. Again, with the help of the editor responsible, we managed to arrange things to our advantage. The two letters raising the most difficult points were suppressed, the one that started off in the most laudatory way was put first, and we wrote a few lines indicating the pettiness and narrow vision of our other critics. So, all in all, this had been a very satisfactory exercise. We had achieved a large amount of publicity without having to discover anything ourselves. I had also satisfactorily demonstrated that, while the pressure of submissions means that *Nature* is obliged often to reject good science, the journal cannot resist a good story.

The Stage

The scientific superstar has two channels of communication with his admiring public. One is regular publication in the fashion journals, and the other is regular appearances at international meetings. Here, the analogy between the scientist and the actor is at its most apt because at the meeting the scientist has the opportunity to present an individual performance, onstage, in front of a large audience, for 30 minutes or more.

There are enormous numbers of scientific meetings because they are organised by all sorts of different bodies such as learned societies, major research institutions, and ad hoc groups who acquire funds from sponsors. Many societies exist only to put on meetings. Many universities and research institutes use meetings to enhance their prestige, either simply by hosting one or by organising one specially to celebrate the opening of a new department or the retirement of a well-known senior figure. In either case the important thing is that international superstars can be got to come to your institution, allowing you to bask in a little of the reflected glory from their brilliance. There has recently been a mushrooming of European-level sponsorship, driven by the availability of money from the European Union to strengthen cultural ties among its members. So if you want to organise a meeting, workshop, or advanced course, just write to the European Molecular Biology Organization, or the European Science Foundation, or Directorate XII of the EU Commission, and you have got your money. Of course, meetings differ in prestige, and the most prestigious are those that restrict entry. These include the Gordon Conferences and Cold Spring Harbor Symposia in the United States, and a few specialist European meetings. Junior staff have to apply for places at such meetings and are really thrilled if they get one because they will then have the opportunity of seeing their favourite stars in the flesh and perhaps even speaking to them. The speakers can acquire some extra prestige by being invited to give a "named plenary." This means the organisers have got the money for your airfare from some company, and in exchange you are giving the "Molecular Pharmaceuticals, Inc. Lecture," preferably at the beginning of the day and to the entire meeting attendance. Although meeting organisers try to get the best stars, they will not always come, and as there are so many meetings going on all the time, even moderately important scientists are in heavy demand to fill all the programme slots.

It can, however, be difficult for junior people to get slots to present their work. Meeting organisers want a good turnout, and they know that this is best achieved by filling the programme with well-known names. Well-known names are guaranteed to give good talks, whereas lesser persons, whom they have heard of only because they have published one or two interesting papers, are something of a gamble. Superstars, and even ordinary stars, need to be booked one to two years in advance because they have a heavy schedule and can be quite picky about what meetings they will attend. With the exception of guest speakers to the "in-house" meetings of companies and large research organisations, speakers at meetings are not usually paid; they just receive their travel, accommodation, and meals. Hence, they are more likely to accept invitations to go to nice places; and for European meetings, it is notable just how many have been switching venue from the cold and rainy centres of science in the north to the warm and vinous Mediterranean coast, or sometimes to attractive resorts in the Alps. After a few years, the novelty of going to meetings wears off, but it remains an obligation to keep going to them because even a moderately successful scientist needs to go on producing results and papers to keep his head above water. Production is mainly done by postdocs, and you are utterly dependent on recruiting good postdocs to carry on your programme. Your chance of getting good postdocs is increased by giving regular good performances at international meetings. They go to see the stars, and if you give a good talk, they may think you are a star. If they are good and come and work for you, they can help make you into a real star. Of course, this isn't the only motivation for going. It has to be admitted that a number of illicit liaisons also occur at meetings, normally of the (male) boss-(female) student variety, and this is an important consideration for some.

So how do you give a good performance? Scientific talks follow a standard format consisting of a lecture illustrated by projection slides. Sometimes people show films, and there are occasional local fashions for running two projectors simultaneously, but the main thing is to be a good lecturer supported by a set of slides. The slides are a crutch without which most scientists will wither and die. They are absolutely obsessive about their slides, making multiple copies, carrying them on their person at all times, and constantly rearranging them in quiet moments when there is nothing better to do. The slides may contain tables of figures or a few sentences of text, or pictures of specimens or of the dreaded gels. Molecular biologists almost never actually measure anything. The results of an experiment are usually presented as a set of gels in which a particular band is more or less strong, showing the presence of more or less of the substance in question. Lecturers often go so fast that it is difficult to remember what kind of gel each slide shows, because there are many different kinds for many different purposes. But nobody can really argue with a pair of nice clean gels, one showing a band and its neighbour not showing it, and the temptation to show photographs of innumerable gels can be compelling.

On the whole, a good lecture is one that people can understand and from which they can remember a single main conclusion. This is what really distinguishes the junior speakers from their more experienced senior colleagues. Students and postdocs invariably think that everyone in the room is an expert, and so they plunge straight into technical details, show slides with huge tables of figures or dozens of gels, and rapidly lose most of the audience. Senior speakers are wise to this, so they tend to throw out all the detailed data slides and substitute slides of particularly beautiful specimens or particularly attractive coloured diagrams. They present lots of background and lots of obvious generalisations. They make sure everyone understands the conclusion right at the beginning and they reemphasise it several times. They also speak clearly and simply and resist the temptation to cover too much material. The result is that they hold the attention of the audience, and when the audience members go home and give meeting reports in their own labs, this talk will be mentioned. This all helps generate the aura needed to attract good postdocs.

Of course, real superstars can also be very tiresome for meeting organisers. The organisers invariably want all the participants to stay for the whole meeting. This will maximise the chances for good discussions, productive interactions, and intellectual synergy. But superstars have so many meetings to go to, and so many named plenaries to deliver, that they tend to want to come just to give their own talk and then go. There are a few well-known American molecular biologists who literally arrive by helicopter to give the talk and then whisk straight off to the next meeting without even swallowing the cocktails or special dinner that adorns the performance. Superstars also tend to incur enormous bills taking taxis from the airport or staying in very expensive hotels, and of course, they expect the organisers to cover any such unforeseen expenditure. Because superstars are at the cutting edge of competitive research, critical data is emerging from their labs every day, and their postdocs will fax or e-mail this new material to whatever meeting their boss happens to be at today. The superstars therefore expect to have continuous access to a fax machine, computer, and slide-making facilities. Finally, superstars are very reluctant to produce manuscripts for those meetings that are financed by publication of a symposium volume.

Symposium volumes deserve a few words if only because they are the lowest form of scientific literature known to mankind, ranking well below primary research papers, reviews, or even single-author books. They exist because meetings are often financed by selling the proceedings to a publisher in order to produce a symposium volume. Publishers like them because, however poor the quality and however outrageous the price of the book, they know they can sell a few hundred copies to the biggest academic libraries in the world and make a guaranteed profit. For such meetings, the speakers have to submit manuscripts vaguely approximating to their talks. Smart meeting organisers never provide any reimbursement of expenses until they have the manuscript in hand, but this rule is not always followed, and once you have given a superstar their money, you can forget about getting your manuscript. Everyone finds the preparation of symposium papers a real bore, and they tend to be thrown together from portions of other papers and contain one or two token pictures. The only people who like producing them are postdocs who want to lengthen their publication lists, but of course, they don't often get asked to speak at meetings.

Meeting of the Rising Sun

The best meeting I ever went to was a symposium on muscle development held a few years ago in Tokyo. This was my first visit to Japan, so I was most interested to see the home of all that is supposed to be good in modern economic and social policy. It was a very lavish affair as it was financed by a Japanese pharmaceutical company. The first unusual feature was the club-class airfare that they provided. Scientists are usually expected to travel by the cheapest possible method, which involves booking particular flights long in advance, staying over inconvenient extra days, and not being able to change your reservations. So British Airways Intercontinental Club Class was quite a revelation to me. You get an enormous seat, rather than the usual sardine size one, you get limitless free food and drink, and even an issue of combs, toothbrushes, and shoe horns in case you might have a sudden desire to put your shoes on. On arrival in Tokyo, we were taken to a large hotel in Shinjuku. Shinjuku is well known as the district that boasts the only skyscrapers in Tokyo. This is because it is the only part of the city built on solid rock and hence fairly resistant to earthquakes. The hotel was extremely lavish, as was the catering for the meeting.

Because I had arrived a day early, and there was not yet the lavish catering, I went out with my colleagues. We wandered around for a bit and realised that it was going to be difficult ordering a meal in a place where all the menus were written in Japanese characters. Although Japanese scientists can be relied on to speak English, this is by no means true of waiters and shop assistants. Because our Japanese consisted of the greeting "hashimemajito" and a few numbers, this was obviously going to be a problem. Eventually, we found a place where the menu contained pictures of several of the items. Knowing that our choice was restricted to these, we pointed. We did fairly well, except for an unforeseen plate of cuttlefish, something I detest and had chosen because the picture had looked more like sliced duck breast.

At this meeting was my colleague Jim Smith, whom we shall meet again in Chapter 5. He had heard that the Japanese were always exchanging business cards. So he had had a set printed carrying a little picture of an embryo and his name, address, e-mail, phone and fax numbers. At that time, an e-mail address was still something of a rarity. I had no cards, and when I saw his, I assumed that I would be at a serious social disadvantage. Fortunately, it turned out that although the Japanese do often exchange cards, they do not expect foreigners to carry them. This meant that there were very few opportunities for handing them out, and by the end of the meeting Jim, who had 100 of them, was reduced to giving them to the European and American speakers who already knew his e-mail address anyway.

It must be admitted that this meeting was of only limited scientific value. But the extremely high quality of catering compensated for this. Each day there was a lavish buffet, with geisha girls in attendance, and all possible combinations of raw fish and vegetables on the table.

One of the British speakers was Tobias Fortune. He is a real superstar but had cultivated a quite different style from those Americans who arrive by helicopter and whose slides are e-mailed to them minutes before they step onto the rostrum. He has made himself the image of the English Country Gentleman. He is courteous, aristocratic, and feigns utter ignorance of any foreign language or custom. He wears tweed suits and would appear naturally at home outside a large country house with his guns and hounds. Foreigners absolutely love all this, and because he also publishes lots of papers in *Nature* and *Cell* and gives good talks at meetings, he is in tremendous demand. He seems to have immense stamina for meetings—at least he is always at the ones I go to—and his name is also always on the programmes of the other international developmental biology meetings that I don't get invited to. He must have given more named plenaries that I have had hot dinners. Although he does thereby attract some excellent post-docs, he also seems to find time to do experiments himself. Truly a re-markable man. Anyway, as the archetypal English Gentleman, he was in-vited to say a few words at the first evening reception. He described a scene from the nineteenth century, shortly after the opening of Japan to foreign trade, in which a number of Samurai had been photographed at the harbour and, fearing that their souls were lost inside the camera, had promptly beheaded the photographer. The grim smiles on the faces of the Japanese suggested to us that this story had been deeply offensive. But for-tunately, the Japanese are very polite, and his English Gentleman aura pro-vided complete protection.

You do have to be careful about your behaviour in Japan. I did remem-ber not to blow my nose in public, which is apparently considered to be approximately equivalent to defaecating on the floor. However, once or twice while I was chairing a session, I pointed at people who had their hands up to speak and then remembered with horror that pointing is very rude in Japan, being almost as bad as blowing your nose. However, the Japanese are very polite to foreign visitors.

After the three-day meeting, Jim Smith, Tobias Fortune, and the other important visitors shot off to their next conference, and I went to Tsukuba to meet Sue Godsave, a former postdoc of mine who was working there. She was quite a quiet and retiring person, and her decision to go to Japan to do her next fellowship was a very brave one. Tsukuba is the so-called Science City, about 30 miles north of Tokyo. It has been growing for about 20 years but still has scientific institutes dotted incongruously around the paddy fields, and there is still almost no public transport from nearby Tokyo. So one of Sue's colleagues kindly drove us there. This and other ex-periences of Japanese road transport persuaded me that when people in Britain talk about traffic jams, they ain't seen nothing yet. People are clearly not going to give up their cars even if things get a whole lot worse than they are at present. I also realised how fortunate we were to have planning regulations that preserve some demarcation between town and country. At a subsequent visit to Hiroshima, in the southern part of the country, I confirmed that the ribbon development does indeed extend along the entire east coast of Japan, and hence covers most of the non-mountainous land in the country. Tokyo itself, like most Japanese cities, has very little open space, and in those days property prices were the high-est in the world. It was sobering to think, when stuck in the traffic jam out-side the Imperial Palace, that this moderate area of green space in the cen-tre of the city was supposed to be worth more than the whole of Florida.

I gave a lecture to Sue's institute in Tsukuba and was impressed by the well-informed grilling I received afterwards. Tsukuba isn't by any stretch of the imagination a "good place," but I found the students and postdocs there rather more stimulating than those in such indubitably good places as UCSF or the Laboratory of Molecular Biology in Cambridge, England. Perhaps it is because they are not worrying all the time about how they are going to get their next paper into *Cell*. I was entertained at a mock Victorian English Tea House, which is really incongruous, situated as it is between the rice fields and the serried ranks of concrete research institutes. Then my hosts insisted on taking me to a karaoke, about which the less said the better.

The day before I was due to return home, they took me to Nikko, which is a really beautiful mountain resort whose surroundings do seem by some miracle to be protected from ribbon development. Nikko has some superb shrines and is the home of the three monkeys: "see no evil, speak no evil, hear no evil." I had long been fascinated by a small wooden carving of these monkeys owned by my father and little thought that I should ever get to see the originals. Nikko also boasts some spectacular waterfalls, and while walking to one of these, we were taken on a very long detour of about three miles to see something really unusual and spectacular. These were enormous fabulous beasts, the like of which were seldom seen by mortal Japanese. We walked along a leafy path towards the strange animals. "There they are," said our host. "Pretty good aren't they?" I looked but could not see anything. "*There*," he said, "Right in front of you." "Ah, yes," I said, feeling rather embarrassed. "Yes, they are pretty spectacular." "Moo," said the strange, half-mythical beasts, which are apparently quite a rarity in Japan. I didn't like to tell him that we could see them out of our bedroom windows at home.

In fact, the really amazing beasts were those in the custody of the Japan Amphibian Laboratory. This is a small private institution that earns its keep mainly by entertaining visiting parties of schoolchildren. Naturally, as I have a professional interest in amphibians, I never pass by the opportunity to visit a collection of them. It was run by the enormously manic Mr. Kokihara, who is a tall, thin, nervous individual I had already met in the Tea House in Tsukuba. He never stopped talking and is a mine of information about the habits of frogs and newts. His star exhibits are a pair of the very rare Japanese Giant Salamanders. They sit pretty still most of the time except for the two occasions per day when a live carp is thrown in their direction. *Snap!* they go, chopping the hapless fish in half with one quick movement. Mr. Kokihara delighted in holding these dragonlike creatures up in the air and allowing them to snap at the schoolchildren with their gigantic jaws and serried ranks of sharp teeth. The children thought it was a great joke.

This, then, had been a good meeting. But in truth, it was not an enormously productive one from a scientific point of view. A good meeting is really supposed to be one at which you find out exactly what all your competitors are doing, arrange to have sent to you any important new clones or antibodies that people have cooked up, and establish several collaborative projects, preferably with "good people" who will get the resulting article into *Cell*. This sort of thing usually happens in America and involves long hours of intense discussion during which you have got to prove that you can remember all the names of all the genes that people have cloned, or expressed, or overexpressed, or knocked out since the last meeting. As I have to tell my wife, each time I set off for the airport: "Darling, it really is not a holiday, it is hard graft."

The Frog and Its Spawn

From the time of Aristotle, biologists have wondered how organisms can develop from eggs or seeds. The children's song

> Oats and beans and barley grow,
> Oats and beans and barley grow,
> But you, nor I, nor anyone know
> How oats and beans and barley grow.

suggests that such knowledge is not for us humble mortals. But academic scientists have large enough egos to think that they can aspire to know even those things formerly reserved for Higher Authority. Since the late nineteenth century, there has been a branch of science called *experimental embryology,* recently renamed *developmental biology,* that has been trying to work it all out. In recent years the egos have triumphed over the eggs, and we now have a good idea in principle of how things work, although many details remain to be filled in.

Eggs and embryos are very small and inconspicuous to the naked eye. So the layman's view of development mainly concerns the growth, or increase in size, that occurs after the stages at which the general anatomy of the body has been formed. For example, as the antiabortionists never tire of pointing out, the human embryo has acquired obviously human features by three months of age, after which its external appearance changes little except for a considerable increase in size. Remarkable as the growth process itself is, the mysteries of development are only fully apparent when you look at an early embryo through a dissecting microscope, or even better, if you watch a time-lapse movie of what is happening. This will show the amazing movements of cells and cell sheets, called *morphogenetic movements,* that transform a simple ball of cells into a complex structure with several cell layers. It will also show the origin of order and pattern, apparently arising spontaneously from nowhere, as the head and tail appear, as particular sorts of cell such as those carrying pigment become visible, and as the heart and blood circulation start up. How on earth does it all work? Can we really understand these wonderful events at the molecular level, in terms of genes and proteins? If so, can we interfere with the process, and, when we reflect more seriously on the implications of interference with embryonic development, should we be allowed to do so?

An explanation of how an egg becomes an animal is obviously not going to be given as a one-sentence answer. The question is complex and multifarious, but the central problem is the one that provided the main motive force for the growth of experimental embryology and that was eventually cracked open in the 1980s by a spectacular three-way combination of old-style experimental embryology, modern molecular biology, and developmental genetics. This was the problem of *regional specification,* or how different body parts are caused to arise from different regions of the early embryo. Why does the head arise at one end and the tail at the other? Why is there only one head rather than two or three? What breaks the symmetry of a spherical egg to produce a bilaterally symmetrical animal? These are the types of issue at stake in the problem of regional specification. Because of the large number of laboratories with their many pairs of hands who have been hard at work since 1980, regional specification is now largely understood. We now know that the origin of pattern arises from particular messenger RNAs deposited in particular regions of the egg. As the egg divides to form separate cells, they each inherit different mRNAs, and this creates the first regional differences between cell populations in the embryo. The pattern then becomes refined because some cell populations emit chemical signals—the inducing factors—that turn on new combinations of genes in the cells that lie nearby, but not in those more distant. There is a complex hierarchy of such processes of signalling and response, which gradually builds up a complex pattern and brings about the regional specification of the entire embryo. For a typical embryo, each major body part becomes specified in terms of a unique combination of active genes at quite an early stage, some time before any visible differentiation takes place. Preeminent in the understanding of the mystery of regional specification have been the contributions of the frog, whose story is told in the present chapter, and the fly, to which we shall return in Chapter 8.

The Frog

People interested in embryonic development have used frog embryos as their experimental material since the late nineteenth century. The reasons are apparent to anyone who gazes into a pond at a mass of frog spawn. The embryos are big. They may not seem very big, but at one to two millimetres in diameter, depending on species, they are very big for eggs. They are big enough that an experienced worker can do quite discriminating surgery on the embryos, without any more elaborate equipment than a few sharp needles and a dissecting microscope. Some readers may be thinking that hens' eggs are even bigger. Indeed they are. The eggs of birds are truly vast for single cells, an ostrich egg measuring perhaps 20 centimetres in length. But in such cases, they are so large that almost all the egg is extraembryonic, consisting not of cells but of food reserves to support embryonic growth. On top of the yolk of a fertilized hen's egg lies the tiny patch of protoplasm that actually becomes the embryo itself. This is no bigger than a frog's egg, and it has already divided to form an embryo of about 60,000 cells by the time the eggs are delivered from the chick factory. This means that some of the early developmental events have already happened, so that the chick embryo, although also a useful experimental material, is not quite as favourable for microsurgery as the embryo of the frog.

Also obvious to the pond-based observer of frog spawn is the sheer quantity of eggs. Compared with most other types of lab animal, frogs are very fecund, and any application that requires a lot of material will benefit from using their eggs. If they look a little closer, pond watchers will notice that the embryos in the spawn are probably at a variety of different developmental stages. This highlights yet another advantage of frog embryos: they are accessible at all stages, from fertilization of the egg to the hatching of the tadpole. Mammalian embryos by contrast are very inaccessible, being buried within the mother for the entire course of development and, for most of it, dependent on the maternal placenta for survival.

Although eggs from a variety of types of frog and newt have been used for embryological investigations, pride of place is nowadays held by a particular species: the African clawed frog *Xenopus laevis* (Fig. 5.1). It has two particular advantages for experimental work that other species do not

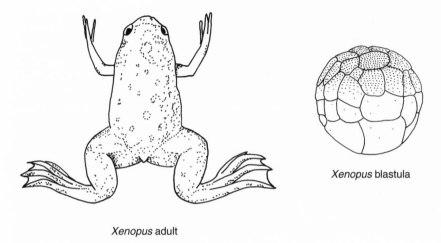

Xenopus blastula

Xenopus adult

Figure 5.1. *Xenopus laevis*.

share. First, it is very easy to keep in the lab, requiring simple tanks, ordinary tap water, and a supply of meat to eat. Second, it is easy to bring about the laying of eggs at any time of the year by injection of a hormone called chorionic gonadotrophin. In the early days of amphibian embryology, workers had to collect their frog or newt eggs from the wild, which necessarily meant that they were only available for a short period each year, during the natural breeding season. It was discovered in the 1930s that injection of pituitary extract could cause a release of eggs (ovulation), but this was very costly in animal lives because several frogs had to be sacrificed to obtain enough pituitaries to cause ovulation in just one recipient. The use of chorionic gonadotrophin overcomes this problem and makes it possible to obtain eggs every day of the year, something that is literally necessary for a developmental biology lab in the competitive times in which we live.

Xenopus originates from southern Africa, where it is very common, often living in the muddy pools at the edge of cattle fields. It was because of its sensitivity to chorionic gonadotrophin that it was first brought in large numbers from Africa into the Northern Hemisphere, being used for the first pregnancy tests during the 1950s. Chorionic gonadotrophin is a hormone made by the early human embryo and subsequently by the placenta, and so it is abundant in the blood and urine of pregnant women. The *Xenopus* test was somewhat more spectacular than the bland do-it-yourself immunoassay kits available from drugstores today. A sample of urine from the patient was injected into some frogs, who were put in a quiet place overnight. If there was enough of the hormone in the urine sample, then the frogs would ovulate. So, the next morning, if the frogs had laid a batch of eggs, then the lady was pregnant. When semipurified chorionic gonadotrophin became commercially available, embryologists started taking advantage of this ability to produce eggs at will, and *Xenopus,* rather than the common frog or newt, gradually came to be the mainline lab animal for studying early embryonic development.

Until recent years many people have been sceptical that one could learn much about the development of humans by studying the embryos of frogs. But this relationship between the reproduction of frog and woman was a useful, if early, indication of the universality of developmental mechanisms. We now know that different animals are remarkably similar in their developmental mechanisms, and that the same genes control development not only in frog and human but even in fruit fly and human. We shall return to this theme in Chapter 8.

The common names for *Xenopus laevis* are the African clawed frog or the African clawed toad. Many people agonise about whether to call it a frog or a toad. Actually, both names are correct because there is no standard zoological definition of what is a frog and what is a toad. It is just a matter of local convention for particular species. For example, in Britain, *Rana temporaria* is known as the frog and *Bufo bufo* is known as the toad, but in other countries, different conventions may apply. Confusion still persists about *Xenopus* because for some inexplicable reason we call them frogs when they are in the lab and toads when they are in the animal house. Perhaps this confusion over nomenclature was devised by the animals themselves to make it hard to keep track of them, for they are great escape artists. If they could read, they would undoubtedly have devoured all the World War II escape classics such as the "Colditz Story." The strong desire to escape means that the first rule of looking after *Xenopus* is to keep the lids on the tanks at all times. If you so much as turn your back on a tank without its lid, then the frogs are hopping around on the floor and making for the nearest dark crevice. The considerable feral population of *Xenopus* in California dates from the time when they were widely kept in hospital laboratories for pregnancy testing. There is also supposed to be a feral population on the coast of South Wales, in Britain. But the climate in this part of the world is less kind to an animal of South African origin than California, and the frogs are often supposed to have died out, until every so often somebody spots another one.

Anatomy

Before embarking on an account of some of the wonderful things that have been done with frog embryos, it is worth mentioning a few anatomi-

Figure 5.2. Axes of an animal.

cal terms that are used every few sentences by those discussing the problems of embryology (Fig. 5.2). The *anterior* of any animal is the head end. The *posterior* is the tail end. The anterior-posterior axis is an imaginary line joining the head and the tail. The *dorsal* side is the side that is normally uppermost and the *ventral* side is the side next to the substratum. These definitions hold for all animals except for that small number of species such as ourselves that walk on two legs, for whom anterior is the same as ventral.

As we saw in Chapter 1, the egg starts its development by cleaving into a hollow ball of cells called the blastula. This then undergoes a complex series of cell movements called *gastrulation,* in the course of which the cells become arranged as three layers. In the frog embryo, this involves cells from the surface invaginating into the interior through an orifice called the *blastopore* (Fig. 5.3). Since the time of Von Baer in the early nineteenth century, it has been conventional to describe the anatomy of animal embryos in terms of three "germ layers" of tissue: the ectoderm, the mesoderm, and the endoderm. The ectoderm lies on the outside and later becomes the skin and the nervous system. The endoderm lies in the centre and later becomes the lining of the gut. The mesoderm lies in between the ectoderm and endoderm and later becomes the muscles, skeleton, kidney, and gonads. During gastrulation, the whole embryo is called a gastrula.

Everyone remembers that Lewis Wolpert said something important about gastrulation, but he didn't remember it himself until I reminded him of the occasion. It was 1979 and we were at a small meeting in Antwerp, Belgium, on the subject of gastrulation. On the first evening, the speakers

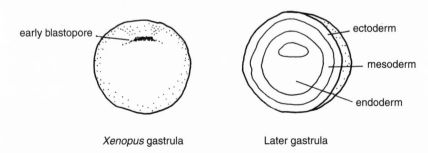

Xenopus gastrula   Later gastrula

Figure 5.3. Gastrulation.

were invited to dinner with various local dignitaries, and Lewis and I found ourselves sitting at a table with a Belgian paediatrician. The organisers had evidently felt that scientists who deal with embryos would have something in common with a doctor who dealt with children, but actually we found the common ground distinctly limited. "What is the meeting about?" the Belgian doctor asked. "It's about gastrulation," we said. "Gastrulation, what is that?" said the doctor. "It's the most important time of your life," said Lewis, without a moment's hesitation. Thus originated the most famous quotation in embryology after William Harvey's "Ex ovo omnia."

Animals are often crudely classified into two groups: the vertebrates and the invertebrates. A vertebrate animal is one with a backbone, such as a fish, frog, bird, or mouse; invertebrate animals are far more diverse and include, for example, flies, worms, and starfish. All animals undergo some sort of gastrulation and the formation of germ layers, but after this the main features of development begin to diverge for vertebrates and invertebrates. In a vertebrate embryo, the next stage of development is called *neurulation,* and the embryo itself is referred to as a *neurula* (Fig. 5.4). During this phase the ectoderm on the upper (i.e., dorsal) side becomes the neural plate, which rolls up into a tube—the neural tube—and sinks below the surface. The main structures after neurulation are also shown in Figure 5.4. The outer covering is the epidermis that later becomes the skin. On the dorsal side is the neural tube that later becomes the brain and spinal cord. These are both derived from the ectoderm of the gastrula. Below the neural tube lies the notochord, a cartilage-like rod that stiffens the early embryo but regresses as the backbone develops. On either side of the neural tube lie segmented structures called somites that form the main body muscles and the backbone. The notochord and somites are both derived from the mesoderm of the gastrula. In embryological jargon, the neural tube, the notochord, and the somites are often referred to collectively as the *axis;* so an embryo that has been modified in some way such

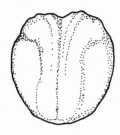

Neural tube closing

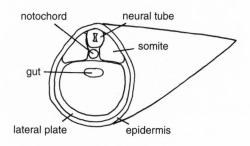

After neural tube closure

Figure 5.4. Neurulation.

that it has a double axis would have two complete sets of neural tube, notochord, and somites.

Hans Spemann, the First Superstar of Embryology

The earliest experiments on amphibian embryos were carried out in the late nineteenth century and involved manoeuvres such as destroying one of the first two cells with a hot needle. In the first decades of the twentieth century, the dominating figure in the field was Hans Spemann, who worked at Würzburg, Rostock, and Berlin before becoming Rektor (i.e., head of the university) and Professor of Zoology at Freiburg. By all accounts, he was a figure of enormous dominance and authority. Photographs convey a rather stern and formal impression, appropriate to a high German official of the period. At the same time, he retained a highly creative mind throughout his career and to the end would engage in animated discussion with members of his lab. He and his students carried out many important experiments over a period of almost 40 years. He was awarded the Nobel Prize for Physiology in 1935 for this work, and by the following year, it was possible for him to write a book giving a reasonably comprehensive account of how the amphibian embryo worked, remembering of course that this was well before the molecular biology era. This book, called in its English version *Embryonic Development and Induction,* is one of the great works of science and is still frequently referred to today, although native speakers of German love to tantalise us Anglo-Saxons by telling us that the English translation is poor and inaccurate.

My German is poor indeed, my embryological vocabulary being just sufficient to read summaries and figure legends in the classical papers. But even this vestigial ability is better than average for a British lab and was sufficient for me to be presented some years ago with a mass of "old German papers" by a colleague with no use for them. When I eventually sorted them out, I found that most of it consisted of manuscript copies of a paper by one of Spemann's colleagues, Alfred Marx, that had been published in 1925. I recognised the paper because by a curious coincidence I had just referred to it in one of my own papers (the one mentioned in Chapter 4 that was rejected by *Nature*). Also among this bundle, I was fortunate to find a handwritten note by Spemann to Marx commenting on one of the drafts. As this was signed at the end, I am now the proud possessor of an original Spemann autograph, albeit one written on ruled exercise book paper.

The essential ideas of Spemann and his school were that each part of the embryo would form a particular structure due to a sequence of chemical signals, or inductions, to which it had been exposed. What a certain part of the embryo would become in normal undisturbed development could be revealed by staining parts of the early embryo with dyes that did not damage it (vital dyes) and then locating the stained part at a later stage. This normal destiny of a part was called the *fate*. But it was realised that the *commitment* of an early embryo part to form the structure in question was not necessarily the same as the fate. In fact, it could be quite different, because if a region was isolated from the embryo at an early stage and allowed to develop in a culture dish, it did not necessarily develop into the same structures that it would have formed had it been left in place. It might form something corresponding to an earlier state of commitment. For example, the part that would normally form the brain does not do so if it is isolated at an early stage, but forms a ball of skin cells instead. It was therefore necessary to distinguish between normal fate and commitment, and so the commitment was called the *state of determination* of the tissue. The nature of determination was the real essence of the regional specification problem and has obsessed most experimental embryologists for generations. What is determination? How does a state of determination become established? Why do different parts of the embryo have different states of determination?

Determination could be investigated experimentally by grafting. If a region of tissue was grafted from one part of the embryo to another, it might form the same structure as it would originally have done, that is, appropriate to its old position, or it might form the structure appropriate to its new position. In the first case, it is said to be *determined,* because it has resisted change; in the second case, it is said to be *undetermined,* because its developmental fate has been redirected by its new surroundings. To interpret this type of experiment, or indeed almost any type of embryological experiment, correctly, it is essential to be able to distinguish the cells of the graft from those of the host. Because grafts heal into place quickly after the operation, and because the shape and size of parts may change a lot due to cell movements, this can be very difficult to do if there is no way of visualising the graft cells. Another great embryologist, the American Ross Harrison, working at Yale University in the early years of the twentieth century, had shown that it was possible to graft tissue between amphibian embryos of different species. Spemann used this method to great advantage as it made it possible to distinguish between the parts of a specimen formed from the tissues of the graft and the parts formed from the host. This could be done, for example, by using donors and hosts that differed in pigmentation. In favourable cases, the pigmentation remained visible for days and was visible not only in the whole specimen but also in the very thin slices (histological sections) that are prepared for microscopical examination to reveal the detailed structure of the cells and tissues. This technique of grafting between species, called *heteroplastic transplantation,* found application in every corner of experimental embryology and remained the most effective way of distinguishing graft from host right up to the introduction of modern cell labelling techniques in the late 1970s.

The heteroplastic transplantation method made it possible to distinguish clearly between a region of the embryo that emitted an inductive signal and other regions that responded to the signal (Fig. 5.5). A signalling centre could be identified in that it altered the developmental fate of surrounding parts of the host embryo into which it was grafted. The induced structure would be an extra copy of the structure that normally developed adjacent to the signalling centre in an undisturbed embryo, as in the middle part of Figure 5.5. On the other hand, a competent, or responding, region of the embryo would be reprogrammed to form something different from its original fate if it was grafted nearer to or further from a signalling centre, as in the lower part of Figure 5.5. It would behave as undetermined, and what it became would depend on the distance from the signalling centre and hence, on the strength of the inductive signal that it received. Nowadays we interpret such results as indicating the existence of some chemical substance emitted from the signalling centre that turns on particular genes in the surrounding cells. We call such substances *inducing factors.* If several different responses are evoked at different concentrations, we call the signal substances *morphogens,* because the signal imparts some pattern, or morphology, to the responding tissue. In Spemann's day the nature of genes was not understood, and it was not usual to think in such strictly mechanistic terms about embryonic behaviour, although, as we shall see, the issue of inducing factors did become quite prominent in the 1930s.

The Organizer

The most famous experiment in all of embryology was carried out in 1921 and 1922 by one of Spemann's students, Hilde Pröscholdt, later to become Hilde Mangold. Because of uncertainty about the origin of the nervous system, Spemann had asked her to carry out a series of grafts in which a part of the gastrula stage embryo adjacent to the early blastopore lip, the dorsal lip, was moved to different positions in a host embryo (Fig. 5.6). When this was done, to everyone's surprise, a second embryo was formed, partly from the graft itself and partly from the tissue immediately surrounding it. The resulting embryos had two complete axes—or sets of neural tube, notochord, and somites—and they bore an obvious resemblance to the "Siamese twins" sometimes born to human mothers. The dor-

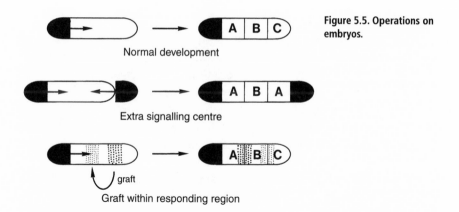

Figure 5.5. Operations on embryos.

Normal development

Extra signalling centre

graft

Graft within responding region

sal lip region became known as the *Organisationzentrum,* or *organizer,* of the early embryo, and has featured in all textbooks of embryology ever since.

The discovery of the organizer was later to earn Spemann his Nobel Prize. Hilde Mangold, alas, did not survive to participate in this honour. Described by Viktor Hamburger as "an unusually gifted, vivacious, and charming young woman," she had married Otto Mangold, one of Spemann's assistants, in 1921. She moved with him to Berlin in early 1924 when he was appointed as Director of the Kaiser Wilhelm Institute for Biology, a position previously occupied by Spemann himself. But on a visit to Otto's family home, she was very badly burned in an accident with a domestic petrol stove and died the next day. This tragedy occurred late in 1924, the same year as the publication of the famous organizer paper, a paper that, almost uniquely for scientific papers from the 1920s, is still heavily cited to this day.

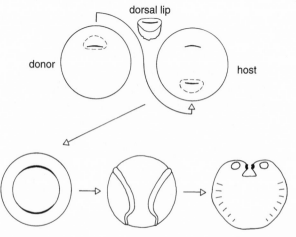

Figure 5.6. The organizer graft.

Everyone in developmental biology has always known that the organizer was important, but what exactly is it and what does it do? It was clear
that the graft of a small piece of tissue from the dorsal lip region would
provoke formation of a secondary axis. In my view, which does of course
have the advantages of hindsight, it was not easy until quite recent times
to decide what this meant, by reading either textbooks or research papers.
There were two major difficulties. First, Spemann repeatedly referred to all
regions of the early embryo other than the organizer as "indifferent," that
is, uncommitted to any particular pathway of development. This made the
signal from the organizer seem quite fantastic. It seemed as though this
small piece of tissue was waving a magic wand to make all the surroundings materialise into the secondary embryo. An associated problem was a
curious lack of precision about the interpretation of the cell origins in the
double embryos. The best of the cases obtained with the heteroplastic
grafting method showed that the notochord of the secondary embryo was
derived from the graft, whereas the neural tube and somites were derived
from the host. This meant that both the neural tube and the somites had
been induced. If the whole embryo was indifferent, then the distinction
between neural tube and somites, the former of ectodermal and the latter
of mesodermal origin, must also be due to the signal from the organizer.
But if the organizer was emitting a chemical signal, then the inductions
might be expected to arise as concentric rings around the graft, corresponding to rings of concentration of the signal. The structure of the secondary embryo did not look anything like a set of concentric rings. For all
these reasons, I do not understand how anyone reading about this subject
before about 1970 can have made much sense of it.

In fact, the organizer experiment was very much ahead of its time, and
its proper significance did not become clear until the 1970s, after another
great embryologist, Pieter Nieuwkoop of the Hubrecht Laboratory in Holland, discovered another, earlier, process resulting in the induction of the
mesoderm. I went to visit Nieuwkoop at his lab in Utrecht in 1977. I had
just started working with early amphibian embryos and felt that I needed
some advice on the techniques of microsurgery and histology. He was very
hospitable and helpful, and I left with what I needed. Nieuwkoop was very
modest and clearly was never once troubled by the fact that his most important articles were not in "good" journals, or that his "impact factor"
could have been better, or that he didn't get enough invitations to important plenaries. He knew that the problems investigated by experimental
embryology were important ones, and that was enough for him. I was very
pleased several years later in 1994 to attend a dinner in his honour at the
Fifth International *Xenopus* Meeting (*Xenopus* was by then important
enough to have its own international meetings). By 1994 he was really fa-

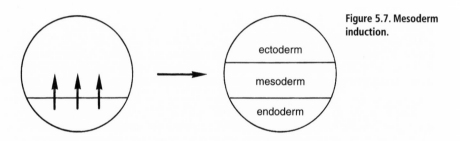

Figure 5.7. Mesoderm induction.

ectoderm

mesoderm

endoderm

mous, at least among those working on *Xenopus,* and the warmth and affection of the hundreds of young researchers for this Grand Old Man were very apparent. He died in 1996, fortunately having seen the problems that had exercised him for most of his working life well on the way to solution. (By the way, my Dutch friends insist that his name should be pronounced "new-kope," not "noy-koop," which is the form particularly favoured by Americans.)

Mesoderm induction is the process of formation of the mesoderm. It occurs before gastrulation and provides the basis on which the organizer works. Briefly, the lower half of the blastula emits a chemical signal that induces a belt of cells around the equator of the blastula to become the mesoderm (Fig. 5.7). This is the signal mimicked by fibroblast growth factor in the experiment described in Chapter 1. So the early gastrula, on which the organizer works, is not indifferent at all. It already consists of three zones of tissue committed to form each of the three germ layers, ectoderm from the top, mesoderm from the middle, and endoderm from the bottom part. The signal from the organizer induces neural tissue from the ectoderm and somite from the mesoderm. It is interesting from the perspective of the sociology of science that the great experiment of Spemann and Mangold, widely recognised as significant and important, did not really make much sense until Nieuwkoop's work of almost 50 years later. To make things even odder, Nieuwkoop's work was itself largely ignored for over 10 years until the work on growth factors had made the field fashionable again.

The First Gold Rush

Notwithstanding the difficulty of interpreting its function in the embryo, it was always clear that the organizer looked like the source of some sort of signal and that it had something to do with the formation of the nervous system, whether directly or indirectly. For this reason, there was great excitement when, in 1932, papers from four German laboratories were published simultaneously in the journal *Naturwissenschaften* showing that *killed* organizer tissue could induce nervous tissue. This proved that the signal must be chemical in nature as it could hardly be a mechanical or electrical effect once the cells were dead. Extensive work by Johannes Holtfreter at the University of Munich showed that inducing activity resided in many tissues of adult animals and could sometimes be unmasked by boiling the tissue. There followed a fascinating period that has been dubbed the "Gold Rush for the Organizer," in which laboratories in Germany, Holland, the United States, and Britain attempted to identify the putative neural inducing factor. It also meant that the problem of thinking about the biological functions of the organizer could be laid to one side while the conceptually simpler problem of the identity of the neural inducing substance took centre stage.

The first tentative identification came just one year later, in 1933, when Spemann, collaborating with the biochemist G. Fischer, identified the substance as the energy storage compound glycogen. Next, a group from Cambridge, composed of the biochemist Joseph Needham and the embryologists Conrad Waddington and Jean Brachet, had a go and concluded that the active principle was an unhydrolysable type of fat, presumably a sterol. Joseph Needham had already made his mark in 1930 by writing a three-volume work entitled *Chemical Embryology*. It may surprise us today that there was enough to say about the biochemistry of development in 1930 to fill even one short article, let alone three large volumes. But Needham had a tendency to the encyclopaedic, which later came to the fore in his monumental series of books about the history of Chinese science. An American lab also investigated the activity of brain extracts and concluded that the active principle lay with another group of fats called cephalins, now known to be among the normal components of the membranes that surround all animal cells. Then the Fischer lab came back with a new theory that the activity was due to certain acidic substances such as thymonucleic acid, a material better known today as DNA.

Things became even more curious when it was found that certain synthetic chemicals had inducing activity. This was true of oleic acid (an unsaturated fatty acid) and of polycyclic hydrocarbons, which were in fashion in the 1930s as they had just been identified as chemical carcinogens, or substances capable of causing cancer. It began to seem as though too many different substances were showing activity and that it would be difficult to decide which of the many candidates was actually doing the job in the embryo itself. The death blow to this line of research came when the Cambridge group showed that a pure synthetic dyestuff, methylene blue, was active. They could put up with materials like the polycyclic hydrocarbons, as it was conceivable that these did exist in embryos and had some physiological function. But living cells, being colourless, obviously did not contain a significant concentration of methylene blue.

By the beginning of World War II, the gold rush had petered out and the principal participants had become disillusioned. Spemann had retired, and died in 1941. Holtfreter left Germany and was interned in Canada for some time before he could return to academic work in the United States. Needham departed for China in 1942 and started his studies of Chinese history that were to occupy him for the rest of his life. The others continued working in biology but on different topics. One of the very last studies of this period, which was ignored at the time but whose significance became clear later, was performed by a Chinese Ph.D. student, H. H. Chuang, in Holtfreter's laboratory. Chuang showed that adult tissues actually contained two distinct inducing activities: a mesoderm inducing activity that was destroyed by heating and a neural inducing activity that was stable to heating. As mentioned in Chapter 1, this study was published in 1939, the same year as the first paper on growth promoting effects of brain extracts, later to become known as fibroblast growth factors (FGFs). But the significance of the mesoderm inducing activity could not really become clear until after the discovery of mesoderm induction by Nieuwkoop.

We now know that the participants in the first gold rush had actually had no chance of isolating any of the inducing factors because the levels of the factors in embryos, or even in most animal tissues, are so low that the technology for their purification was not available until the mid-1980s. That different laboratories had identified different classes of substance as the active principle showed, with the benefit of hindsight, only that the real activity was a minor contaminant of their crude preparations. But the results of the gold rush had a very negative effect on inducing factor research. The idea became firmly established that the induction of nervous tissue, or indeed the induction of anything, at any stage of development, in any organism, was a nonspecific response that could be triggered by a wide variety of substances. The idea that inducing factors were nonspecific and uninteresting continued to haunt the biological sciences for decades, so much so that when it became fashionable again in the 1980s, workers were really surprised when they found that it was not true. At least when one is working with *Xenopus,* it is quite hard to find active inducing substances, and most of those that are now known probably are bound up in some way with the actual endogenous mechanism.

The Interregnum

Spemann himself died in 1941. Because of World War II, experimental embryology almost died with him. In the interwar years, it had been a very German subject. Many British and American scientists had travelled to Germany to work, they read the latest papers in *Wilhelm Rouxs Archives,* and some even felt impelled to publish in German themselves. Of course, the war changed all that. German research came to an abrupt end, and the scientists were killed or dispersed, some ending up in the United States. By the time that German science had been rebuilt, embryology had fallen out of fashion. An indication of its unfashionable status is that the second Nobel Prize for an achievement in developmental biology was not awarded until 1995. It was given to three workers who had effectively solved the molecular biology of the development of the fruit fly *Drosophila* (see Chapter 8). One of the recipients was Christiane Nüsslein-Volhard from Tübingen, showing that German developmental biology had once again become important. But by now the United States has been dominant in science for so long that her laboratory has to publish and operate in English in order to have an impact on the international stage.

As mentioned above, the Cambridge group active in the days of the gold rush was that of Waddington, Needham, and Brachet. Waddington was Conrad Waddington, whose book *Principles of Embryology* was such a strong influence on me when I entered developmental biology (see Chapter 2). Needham was Joseph Needham. As previously mentioned, he was later to become famous for his research on the history of Chinese science and technology and for his endless multivolume work on the subject. Brachet was the Jean Brachet we have already met in Chapter 2. The first two I saw only once. I saw Waddington at a seminar in Edinburgh on the effects of a carcinogenic dyestuff, trypan blue, on the development of lizard limbs. I saw Needham in Oxford at a lecture on Taoist alchemy and medicinal chemistry, from which it appeared that the ancient Taoists had discovered just about everything worth knowing. But Brachet I encountered several times on various visits to Brussels. He delighted in recounting stories of the old days to incredulous younger listeners. He spoke English very well, but with a strong Belgian accent that gave his stories an extra piquancy. His father had also been an embryologist of renown and had taught his son the intricacies of microsurgery on frog embryos in the 1920s. The technology for operating on frog embryos has not changed much over the years and still involves freehand dissection with sharp needles under the dissecting microscope. But back in the 1920s, life was much harder for the embryologist as Brachet explained in one after dinner speech to a youthful audience. In the 1920s and 1930s, embryologists had to collect their eggs from the wild, which meant that the experimental season was limited to about three weeks in the spring. Also, there were no antibiotics, so almost all of the operated specimens became infected by bacteria and died before the results of the experiment were apparent. For example, Hilde Mangold had performed 259 heteroplastic organizer grafts in 1921 and 1922, but only 26 survived, and only 6 were deemed good enough to be included in the famous paper. Apparently, Brachet senior had one day turned to his son and said wistfully, "If only we had a refrigerator." His son replied, "A refrigerator, how wonderful! We could keep the frogs in it and extend the breeding season over many more weeks!" Brachet senior sadly said, "Out of the question. We just don't have that sort of money." The young scientists listening to this of course all came from labs stuffed to the ceiling with refrigerators, freezers, incubators, and hundreds of other pieces of equipment. They listened, minds blown, trying hard to imagine how anyone could have discovered anything in the past.

In the period from 1950 to 1980, experimental embryology existed in a curious sort of twilight world. It was not that nobody was working on embryological problems. They were, and in larger numbers than in the heroic days of the 1920s and 1930s. The induction issue was kept alive by a very small group of whom the Finnish embryologists Lauri Saxen and Sulo Toivonen were preeminent. But biological sciences had entered the phase of exponential growth, so the proportion of those interested in embryos had diminished and their profile in terms of publication in "good" journals and plenary lectures at big meetings was relatively low. Moreover, in 1953 the structure of DNA was published, unleashing the explosive growth of molecular biology that was eventually to leave no corner of the biological sciences unscathed. In a curious way, the solution of the great problem of the mechanism of inheritance had made people forget that there was also a problem of the mechanism of development. Inheritance had been the central mystery of biology since the time of Darwin and Mendel, and by the 1950s it had been solved. Genes were composed of DNA, they worked by directing the synthesis of proteins, and the proteins did all the biochemical jobs like enzymic catalysis, membrane transport, and maintaining the structure of the cell. It therefore seemed to most biologists that the question "How do embryos develop?" was a nonquestion and should be reduced to the simpler one: "How is gene activity regulated?"

Nuclear Totipotency

It is probably for this reason that the one embryological problem that did gain wide publicity in this period was the issue of whether different sorts of cell lost subsets of genes or retained them all. It was clear that cell types like nerve, muscle, or skin were different from each other because they were composed of different types of protein and were therefore expressing different genes. If all genes were retained in all cells, then there must be special mechanisms for turning genes on and off in order to produce the different cell types. Once again the frog came to prominence to help answer this question. In 1952 two American embryologists, Robert Briggs and Thomas King, working at the Institute of Cancer Research in Philadelphia, showed that it was possible to transplant a cell nucleus from the multicellular blastula stage of development back into the egg. If the egg's own nucleus had been destroyed by ultraviolet irradiation, the donor nucleus could step into its place and support the development of the egg. If the donor and host came from different genetic strains, differing for example in the skin pigmentation pattern, then the embryos developing from the nuclear transplantations had the attributes of the nucleus donor, rather than those of the egg donor. This experiment confirmed that the genes were indeed in the nucleus, showed that no genes were lost at least as far as the blastula stage of development, and also showed that it was possible to "clone" animals.

A *clone* is something created by repeated asexual replication, for example, a colony of cells grown from a single cell, or a piece of DNA grown up in a bacterial host. Animals with identical genetic constitution, such as identical twins, are also "clones" of each other. Briggs and King had shown that it was possible in principle to create a large number of genetically identical frogs by transplanting nuclei from a single embryo into many host eggs. This result gripped the minds of journalists and science-fiction writers, and ever since we have regularly been regaled with newspaper stories about the cloning of new individuals from tissue fragments of ancient Egyptian mummies, Hitler's corpse, aliens, and so on.

Several other workers who were interested to know whether nuclei from fully differentiated cells could support development of an egg continued studying the potential of nuclear transplantation. The best-known experiments were those performed by John Gurdon, first at Oxford and later at Cambridge. Gurdon is an embryologist of great international renown who has received over 30 medals and other awards for his work. It is interesting that he was not permitted to study biology at his high school, the expensive and exclusive Eton College. In fact, his biology teacher declared at an early stage that "Gurdon is the worst pupil it has been my lot to teach," and he was obliged thereafter to study Latin and Greek. Fortunately for science, he was able to return to biology at Oxford University, and during his Ph.D. work, he initiated a long series of further studies on nuclear transplantation using *Xenopus* rather than the leopard frogs of Briggs and King. Eventually, he showed that it was possible to support development of an entire embryo to an early tadpole stage using nuclei from cultured skin cells. Although this worked, it did so with a very low yield. In other words, only a tiny proportion of the transplants survived and developed, and most of these were abnormal in some way. This experiment is interesting for sociologists of science because such a low yield in an embryological experiment would normally have been counted as a negative result. In this case, however, the small number of positive cases were accepted not only as positive but also as decisive and definitive. This is partly because the result was the expected one, but also because there are many possible ways in which a nucleus could fail to adapt to the rapid division rate characteristic of the early embryo. This experiment has featured in the first chapter of textbooks of developmental biology ever since as evidence that all genes are retained in all cell types of all animals (or almost all, as one can never really say "all" in biology; for example, there are now known to be DNA rearrangements in the B and T cells of the immune system).

For various reasons it had been supposed that it would be more difficult to do the same sort of thing in mammals. But this may not be true, because tissue culture technology is much better developed for mammals than for amphibians. In 1997 a group in Edinburgh reported the successful cloning of a lamb by the fusion of a cell cultured from an adult ewe with an oocyte whose own nucleus had been removed. This story created enormous excitement in the media and led to a panic about the possibility of human cloning. Not for the first time, something that had started in the frog proved in the end to apply to mammals as well.

Gradients and Factors

During the interregnum interest in induction and embryonic patterning had fallen off, but activity continued in various nooks and crannies. Saxen and Toivonen in Finland published some models that involved hypothetical gradients of factors responsible for patterning the early amphibian embryo. There was also a long saga of protein purification by the German biochemist Heinz Tiedemann, working in Berlin, who eventually succeeded in purifying the mesoderm inducing factor (called by him the *vegetalizing factor*) to homogeneity. After having read about the gold rush, readers may well be surprised that Tiedemann did not receive a Nobel Prize. But he had a number of things stacked against him. First, as I am sure he would be the first to admit, he is not an individual of high charisma. In an atmosphere in which embryology had fallen out of fashion and the hunt for inducing factors was regarded as hopeless because of nonspecificity, attention would only have been paid to a spectacular platform performer. Regrettably, poor Tiedemann was even nicknamed "Herr Doktor Tedious" by some irreverent Anglo-Saxons. Second, he obtained the factor from late chick embryos, not from amphibian embryos themselves. This was essential because amphibian embryos can only be obtained in gram quantities, and to purify substances of such low abundance as growth factors (as we now know these substances to be), kilogram quantities of tissue are required. Third, the factor was not purified for many years, and with each successive paper, the purification protocol became lengthened by another step until it was very long indeed. Because the work took so long, the first steps were beginning to look positively antiquated by the time that the protein was approaching purity. Finally, the work was begun in the BC era and finished after cloning had become well established. By then nobody would believe in the reality of a new protein just because it had been purified! A clone and an amino acid sequence were now required, and these were not obtained until it was too late (see below). I visited Tiedemann in Berlin in the early 1980s. He had a large team to assist him in the purification of the vegetalizing factor, but perhaps the lavish resources had held back his creativity, as he was still using very labour-intensive and antiquated methods. Still, he was kind enough to send me a sample of his factor when I later wrote to request one. This had quite an impact in my lab and is the moment from which I personally felt the onset of a second "gold rush."

The Second Gold Rush

Unlike me, whose pulse quickened at the sight of the inductions provoked by Tiedemann's sample, history will probably date the commencement of the second gold rush for inducing factors with the discovery of mesoderm induction by Pieter Nieuwkoop. As mentioned previously, mesoderm induction is the first inductive interaction in the amphibian embryo. Cells in the lower, vegetal, hemisphere of the embryo emit signals that induce a ring of tissue around the equator of the blastula to become the mesoderm, that is, to develop into muscle, connective tissue, and other characteristically mesodermal tissue types (Fig. 5.7). The dorsalmost part of the mesodermal ring is the organizer, and it acts both on the rest of the mesoderm, to induce the somites, and on the ectoderm, to induce the neural plate.

At this time I had a very good postdoc in my lab named Jim Smith. He had done a Ph.D. in Lewis Wolpert's lab, and then a postdoctoral period working with a growth factor (PDGF) in the United States. So he had exactly the right background to make an impact on the induction problem. His project in my lab had been to repeat the organizer experiment using one of the new cell labels that had just been introduced in the United States. In the old days, Harrison and Spemann had used grafts between different species in order to distinguish which parts of the resulting embryo came from the graft and which from the host. But this method never really allowed high resolution discrimination. The new labels offered the chance of getting much more precise answers. This study, done on *Xenopus* and including many more cases than the original organizer paper, had confirmed its essential findings, and, in particular, confirmed that the graft induced somites from the mesoderm of the host, as well as a neural tube from the ectoderm. It seemed to us that the organizer signal acted to "dorsalize" the embryo, because neural tube is dorsal ectoderm and somite is dorsal mesoderm. This aspect of organizer function had not really been clearly recognised before as most textbooks of developmental biology at the time presented the organizer just as a neural inducing agent.

Although our paper was just a minor footnote to Spemann and Mangold, it was satisfying to know firsthand what did happen following an organizer graft, and to know that it worked in *Xenopus* as well as the various wild-caught newts and frogs available in the 1920s. This study of embryonic induction made Jim very interested in the nature of inducing signals, and he was particularly interested when I told him about Tiedemann's vegetalizing factor, and by the demonstrable activity of the partially purified sample from Berlin. We decided to try to purify the factor ourselves, using a different approach from Tiedemann. One of the problems with Tiedemann's work was that the factor was prepared in an insoluble form and had to be assayed by grafting pellets into embryos. As with the BMP assay discussed in Chapter 3, this method could almost have been designed to send shivers down the spines of biochemists. If something is not in solution, it is not possible to measure its concentration; if you can't measure the concentration, it is not possible to monitor the progress of a purification. So we started off by developing a new assay that did work in solution and allowed some sort of quantitative measurement of the amount of activity in a sample. But things slowed down after this advance, as the purification work proved very difficult. At this time, Jim's three-year fellowship was up; so he left my lab and moved to a tenure-track position at the National Institute for Medical Research. His growth factor background had made him aware of the advantages of using tissue culture cells as sources of growth factors. The main advantage of cells over animal tissues or embryos is that the factor is secreted into the medium, so it comes already in solution and does not need to be separated from all the cellular protein. He soon found a *Xenopus* tissue culture cell line (XTC cells) that secreted a mesoderm inducing factor, and this proved to be a more favourable source than the chick embryos. Because my purification work was so slow, and because I did not want to compete too directly with my former postdoc, I eventually abandoned the purification of the chick factor and resorted to using our assay to test a number of candidate factors, hence leading to the experiment described in Chapter 1.

Jim published a preliminary account of his factor in January 1987; we published our result that FGF was a mesoderm inducing factor in March of the same year. In December David Kimelman and Marc Kirschner from UCSF published another paper very similar to ours about FGF being a mesoderm inducing factor. I was astonished by this, because having spent some years in a scientific backwater, I was not used to the fast lane of competitive science. You can tell all too easily when you are in the fast lane because several labs often end up publishing an important result at the same time. In retrospect I realised that UCSF was in 1986 one of the few other places in the world where *Xenopus* embryos and growth factors could both found under the same roof. Marc Kirschner is an almost legendary hero of cell biology. He is a large individual in all dimensions and was doing pioneering work on cell division and the dynamics of the cytoskeleton at the same time as the mesoderm induction work. He has a limitless ability at scientific meetings to discuss complex technical issues throughout the night. He had become acquainted with the problem of mesoderm induction during a sabbatical spent in Nieuwkoop's institute in Holland, and given the interest in FGF at UCSF, it was only a matter of time before the factor and the embryo came together. David Kimelman was the junior author at the time, but he has since led many fine studies of his own on early development. Next to their paper (in the journal *Cell*) was published a study from the lab of another American, Doug Melton from Harvard University, showing that there was a messenger RNA localised in the lower hemisphere of the *Xenopus* egg, where the mesoderm inducing factor should be, and that its amino acid sequence resembled another known growth factor, transforming growth factor beta (TGFβ). This TGFβ-related factor was called vg-1 (pronounced "vedge-one") because it was the first mRNA to be discovered to be localised to the *vegetal* hemisphere.

Doug Melton is almost the polar opposite of Marc Kirschner. He is small, dark, and slim. Very unusual for star scientists, he dislikes travelling and is quite hard to lure to a meeting. He is laconic in his conversation but can move very fast indeed in his research. Once he has decided that a problem is ripe for solution, woe betide the competitor who dares to race him to the finishing post. Melton had approached the problem of mesoderm induction from a quite different direction, looking for any mRNA whose abundance differed in the upper and lower parts of the egg. His search had yielded several mRNAs enriched in the upper hemisphere, but only one in the lower hemisphere, and this was vg-1. So in the end, all of us—Jim, myself, Kimelman, and Melton—had approached the problem in different ways and had published our papers in the same year. Clearly, the identification of inducing factors was an idea whose time had come, or in view of the first gold rush, perhaps one should say, "come again."

The Molecular Follow-Up

What were we to do now? A friend of mine whom I had met in America that summer said prophetically: "You will have to decide whether to follow the factor or the biology." Indeed, modern developmental biologists seem divided into a majority who are interested in a particular gene or gene product ("I'm studying the functions of FGF") and a minority studying biological problems without preconceptions about which molecules are important in this context ("I'm studying the development of the liver"). A significant investment of time is needed to study any particular gene because of the various complex clones and probes that need to be prepared. This creates a strong temptation to specialise in a molecule, but some of us have always felt that the biological questions should, if possible, be kept in the forefront.

Nowadays it is essential to clone the gene for any protein of potential interest from the organism under study, in our case *Xenopus*. The clone is then used to make a probe to study the *expression pattern* of the gene—in other words, the developmental stages and the regions of the embryo in which the gene is active and hence making mRNA. The expression pattern is the first thing you need to know in order to assess the biological function of the gene. Clearly, if the gene is not expressed in a particular place, then it is not doing anything there. If it is expressed, then it may be doing something, although this is not necessarily proof of function: even if the gene is expressed, the product may be inactive for some reason. To positively establish function, more data is needed than just the expression pattern. In particular, one needs to know the effects of overexpressing the gene in the embryo and also of inhibiting its normal activity in the embryo.

The technique of overexpression of genes is a particularly useful one in developmental biology. The method used for frog embryos had been invented by Doug Melton and his coworkers in 1984 and consists of making an RNA copy of the cloned gene in the test tube and injecting this RNA into one or more cells of the embryo. Once inside the cells, it becomes translated into protein, and the biological activity of the protein may bring about some visible change in the anatomy of the embryo. The inhibition of function of a particular gene is often studied by means of genetic mutations in flies and mice, but this option is not possible in frogs because of their long life cycle. Instead, it is necessary to design a specific inhibitor for the gene in question, make the mRNA for this in vitro, and then inject the RNA coding for the inhibitor into the embryo. This has also proved to be a very important technique in recent years.

Based on what we knew of its chemical properties, it was clear even in the early days that Jim's XTC factor was not the same as FGF. Three years later he was able to identify it as another already-known growth factor called *activin,* which, like vg-1, is related to the TGFβ factors described in Chapter 3. It later turned out that the vegetalizing factor of Tiedemann, the factor that in some ways had started it all, was also activin. There then followed a race between Jim, Doug Melton, and Igor Dawid at the National Institutes of Health in Maryland, to clone activins from *Xenopus.* Melton won, as he usually does, but the resulting experiments showed that the activin gene was not expressed in the early embryo at the time when mesoderm induction is going on, so activin itself could not be the endogenous mesoderm inducing factor.

The final outcome of the second gold rush is still not agreed upon by all, but there is some consensus that vg-1 is the "true" mesoderm inducer because it is found in the vegetal cells, it has the appropriate activity, and when it is inhibited, the embryo forms no mesoderm. Meanwhile, several laboratories, including mine, have shown that FGFs are, contrary to expectation, present in the upper hemisphere of the blastula (i.e., in the responding cells rather than the signalling cells). The FGF is not at sufficient concentration to induce mesoderm in the animal hemisphere, but it does stimulate the cells to some degree, and this stimulation is necessary for them to respond to vg-1 (Fig. 5.8). If the FGFs in the early embryo are inhibited, the cells of the upper hemisphere are unable to respond to the vg-1 signal. So in the end it seems that both factors are necessary for mesoderm induction, and as often happens in science, all the labs involved can be happy that they have contributed at least a piece of the jigsaw to the final solution.

Molecular Basis of the Organizer

All the excitement associated with the second gold rush has more or less solved the problem of the mechanism of mesoderm induction. But what of the original organizer phenomenon, the induction of a second embryo following the implantation of a dorsal lip on the ventral side? By now it had become clear that we needed to explain two distinct things: first, how the organizer itself is formed in development, and second, the nature of the signals emitted by it.

The solution to the first problem was suggested in 1989 when two American groups overexpressed a Wnt gene in *Xenopus* embryos. As mentioned in Chapter 3, the Wnt factors had been discovered both by cancer researchers in mice and by those working on the development of the fruit fly *Drosophila.* When a Wnt gene from mouse or fly was overexpressed in *Xenopus,* it caused the formation of a double axis, rather reminiscent of the

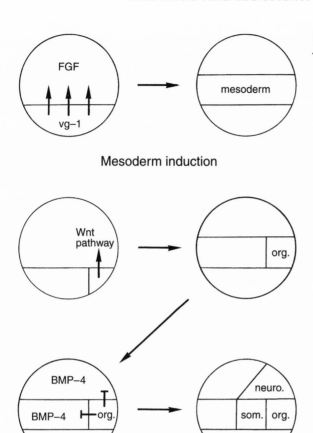

Figure 5.8. Inductions in *Xenopus* development.

Mesoderm induction

Dorsalization

original organizer graft. Once again, because it was thought that such effects were nonspecific, there was no immediate follow-up. But a few years later, two papers appeared simultaneously that started the "Wnt Rush." One was from Melton, who had cloned a Wnt gene from *Xenopus* and showed that this, like the fly and mouse genes, also caused double axis formation. Another was from Richard Harland, from the University of California at Berkeley, who had developed a novel method for screening large numbers of unknown clones for biological activity. Richard Harland is a tall bluff Englishman from the North Country, where folk are well known to be straightforward and true, without frills or nonsense. He preserves this simple exterior, but like the other stars, he can be very efficient when it comes to moving fast in research. His screening method has produced several important genes, and the first one was a Wnt gene. Subsequently, many laboratories have examined the biochemistry of Wnt action in great detail, and it now seems clear that the organizer in *Xenopus* arises because of the activity of a Wnt-like factor on the future dorsal side. This cooperates with the vg-1/FGF signal to turn on several genes on the dorsal side of the embryo, and it is the activity of these genes that gives the organizer region its characteristic properties (Fig. 5.8).

As we have seen, the biological action of the organizer is to induce somite from the mesoderm and neural plate from the ectoderm. Both of these actions can be regarded as *dorsalizations,* because the tissues are caused to form more dorsal-type structures than they would in the absence of treatment. It turns out that there are substances secreted by the organizer all with a similar mode of action. What they actually do is antagonise a ventralizing substance called *bone morphogenetic protein-4* (BMP-4). This factor, described in Chapter 3, is one of the BMP family of growth factors and was first discovered as a substance promoting bone development in rodents. It is expressed in early *Xenopus* embryos and is now known to be necessary for ventral type development, the formation of epidermis from the ectoderm or of blood cells from the mesoderm. The organizer region secretes several factors that bind to and inactivate the BMP-4 protein. Thus, in the region near the organizer, ventral development is suppressed and the alternative, dorsal, programme is followed, causing the ectoderm to form a neural tube and the mesoderm to form somites (Fig. 5.8). The organizer therefore works by the inhibition of ventral development rather than by the promotion of dorsal development. To put it another way, it is an inhibitor of an inhibitor, and two inhibitions are together equivalent to one activation.

The final aspect of organizer function is the formation of the anterior-posterior, or head-to-tail pattern of the body, as the induced axis has an anteroposterior pattern like the host axis. This is a complex ongoing area of research. But it is of special interest to me because it involves our friends the FGFs. There are in fact many different kinds of FGF, and since the days of the experiment described in Chapter 1, my lab has been studying the endogenous FGFs of *Xenopus*. As we have seen, some FGFs are expressed in the upper half of the blastula and are necessary for mesoderm induction at the blastula stage. But we have also found that several of the FGFs are expressed at the posterior end of the embryo during gastrulation. It is now known that the anterior-posterior pattern of structures in all animals is coded by the expression of a group of genes called the Hox genes (which we shall meet again in Chapter 8), and the function of the FGFs in the frog gastrula seems to be to turn on those Hox genes required for the formation of the trunk-tail part of the body. So if FGF signalling is inhibited, the embryo develops as an isolated head, with no trunk or tail. For this reason, we now believe that the FGFs are not only needed for mesoderm induction but are also an essential component of the mechanism for anterior-posterior patterning.

Full Marks for the Frog

This account has only covered some of the contributions that the frog embryo has made to our knowledge of development, and it is heavily biased by my own interest in the mechanism of induction in the early embryo. But it shows the enormous progress in our understanding that has come from the fusion of classical experimental embryology with the modern technology of molecular biology. Curiously enough, those of us who work on frog embryos have often been on the defensive in the developmental biology community. We are often reminded that frogs are not suitable for genetic experiments because of their long life cycle. In this respect they contrast dramatically with the fruit fly *Drosophila* that sometimes seems to have been put on the earth by the creator specifically for the benefit of experimental geneticists. The contribution of *Drosophila* will be discussed later, but the truth is that there is an enormous synergism between different lines of work. The functions of genes discovered in *Drosophila* can often be studied more easily in *Xenopus*. Growth factors isolated by cancer researchers can be studied at the level of biological functions by developmental biologists. Molecular and microsurgical techniques combined produce more than either on its own. Above all, work on frog embryos does not tell us just about frogs! We now know that the genes that control development in all animals are very similar and that the results from frogs apply in large measure also to mammals, including the economically important farm animals and ultimately including also ourselves.

Who's Who in the Lab

A peculiarity of science as a profession is that the formal qualifications of a first degree and a Ph.D., although they are hurdles to be jumped, do not in any way guarantee a future career. In this respect it is quite different from traditional professions such as law, medicine, or accountancy, where the possession of the qualifications will virtually guarantee some sort of employment in the future. Science is not like a traditional profession at all. It is more like acting, with its emphasis on individual virtuosity, the overriding importance of public popularity, and the high drop-out rate. In the early years, the positions are all temporary, usually for a maximum of three years. Because of the relatively large amounts of grant money in the system, there are many temporary graduate and postdoctoral assistantships available. But few people from this stage manage to make it to one of the coveted positions permitting scientific independence, meaning the possibility of choosing your own problems to work on and having a lab under your own authority. Even better is to have tenure, which means you are not only independent but also have a permanent job, even if it is not quite as permanent as it used to be. The main routes through the system are depicted in Figure 6.1, which shows the positions generally found in the United Kingdom. The U.S. system differs in detail but not in general conception.

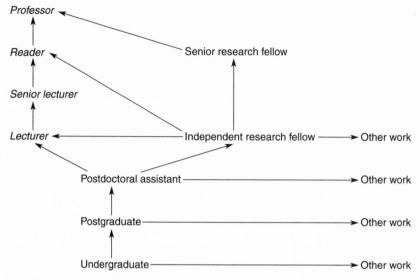

Figure 6.1. Career paths (U.K. version). Italicised posts are staff positions in universities or their equivalent in research institutes. Only these are tenured.

Here is a list of the people you might meet if you were to visit a lab:

The Director. Head of an institute or special unit. He, or occasionally she, is important because he has control over new appointments and over the use of space. But he doesn't have much time for his own research and may lose street credibility because of his distance from the bench. He has tenure.

The Professor. Commonly thought to be of great eminence, but now just a senior faculty member in a university. In the United States everyone is a professor; in the United Kingdom most departments have several professors, but the professor who is chairman, or head of the department, is more important than the others because he has control over space and appointments. Professors normally have tenure.

The Lab Head (Principal Investigator). In a university the lab head could be any grade of academic. In a research institute he would have some title such as "senior scientist." He controls the research of one lab, with its associated fund-raising and recruitment. Commonly known as "the boss." He probably has tenure, but may be "tenure track," which means that he is allowed to apply for tenure in due course.

The Postdoc. Nowadays science is so complicated that the mere possession of a Ph.D. degree is not thought to equip you for much. A serious candidate for a career in science is expected to do a couple of postdoctoral fellowships or assistantships after receiving the Ph.D. These are usually each of two or three years fixed-term duration, after which the postdoc has to find a "real job."

The Student (Ph.D. Student, Graduate Student). In Britain they are science graduates who work in the lab for three years then write up what they have been doing as a Ph.D. thesis. They are usually paid a pittance. In the United States they would normally get one to two years of more formal course work or attachments to different labs in addition to the main research period. But they are still paid a pittance.

The Technician. In some places, almost an extinct species.

The Research Assistant/Research Officer/Scientific Officer/ Experimental Officer. Formerly known as technicians, but can do any job in the lab ranging from the essential but unglamorous routine maintenance tasks to the higher-status but often less productive experimental work. What they do really depends on what has to be done and what they, as individuals, are good at. In institutes they often have permanent jobs, but in universities many are grant-supported, and like postdoctoral assistants, they expire with the grant.

The Lab Head

Once we were sitting in the pub having a laboratory Christmas outing. "You know," said Betsy, "You can always tell how important a scientist is from the way he dresses." It's true that department heads and professors sometimes wear jackets and ties whereas the more junior people wear jeans and sweaters. But the really important scientists wear tee shirts and jeans that are old, torn, and dirty. They even wear their old, dirty, and smelly clothes when they are giving prestigious public lectures. They are making a very serious statement. *I don't need to dress up for you. I am brilliant. I publish in Cell. You need me more than I need you, so I shall wear what I like.*

This observation contains a key truth unknown to the outside world and often recognised only implicitly by scientists themselves. The person who has authority over hiring, space, and money, such as a department head in a university, commands considerable respect from secretaries and technicians. But he may command little respect from other scientists. The temporal powers wielded by such people are of strictly local significance, and there may well be other scientists in the department who are better known externally. This external reputation is critical because the health of the department or institute depends directly on it. If there are well-known stars around, they act as magnets for others, both junior and senior, who will want to come and work nearby. They will be able to attract large research grants, which will keep the department afloat financially. Most of all, they will contribute to a general consensus that the place in which they reside is a "good place."

These stars are simply independent lab heads, and the main job of the lab head is to get the money to hire the members of his lab, which is usually in the form of grants from various external funding bodies. Getting them is exceedingly competitive and involves much effort in the preparation of applications that promise innovative, important, and successful research. Once he, or sometimes she, has got the money, he must appoint his staff, train them in the required techniques, supervise their work, overcome the inevitable many failures and difficulties, and make sure it gets written up and published in the highest-impact journals possible. The management tasks will include dealing with the very complex regulations about chemical, radiation, or biohazard safety, and animal use in his lab. He should also regularly give lectures in other institutions and attend scientific meetings to present the work of the lab. He must be aware of all work in the field worldwide, and it helps to know most of the competition personally. He will doubtless sit on various committees outside his institution and do a large amount of refereeing of scientific manuscripts and grant proposals. Considering that they also usually have to do teaching and some departmental administration, lab heads nowadays often have little time left for actually doing experiments themselves. It is a sobering thought that "research" to an arts academic means reading things in the library. But this is what the scientist has to do in order to have the ideas to put in his grant application, before he has started any of the actual work!

The Postdoc

What you can do in your lab depends very largely on who you have working with you, so recruitment is of enormous importance. Although lab heads have typically worked very hard in their own junior days, by the time they have become lab heads, indolence and the pressures of family life have usually set in, so their working hours are reduced to 50 to 60 per week, they often don't spend evenings and weekends in the lab, and worst of all, their daily diet of supervision, administration, fund-raising, committee meetings, teaching, and recruitment use up so much of their time, they cannot do full-scale projects any more and are only good for the odd pilot experiment.

So what constitutes a good assistant? It doesn't actually matter whether an individual comes in as a technician, a Ph.D. student, or a postdoc:

what they do depends on their individual capabilities. A good person is, naturally enough, one who understands the nature of the lab's work, who has ideas of their own, can get experiments to work, and can finish things as well as start them. It is also helpful if they have pleasant personalities and are always smiling and being nice to people. This is desirable in the cramped conditions of a lab but is often incompatible with getting experiments to work, because success in this art requires a totally obsessive attitude to detail and a deeply pessimistic outlook in anticipating possible problems.

But you don't want your assistants to be too good, or you start running into another type of problem. Good people are always on the point of leaving because they keep being offered jobs elsewhere. In science a good place has to work hard to stay good. It can only do so by recruiting good people, and as we have seen, good people are easily identified because they publish their work in good journals. This is a particular problem for the department chairman, or institute director, since the good lab heads are always hinting that they are about to leave, although they might possibly be induced to stay if more space or resources were pushed in their direction.

But it is even a problem for a lab head: if fortunate enough to get a good postdoc, that person doesn't usually want to do the project laid out for them. After all, being good, such postdocs have their own ideas about what is a good project. They also tend to open up collaborations with good people in other places. This is an efficient way of publishing more papers in good journals in a limited amount of time, but the lab heads may get frozen out. Then, you do not know whether they will be there in three months time, as they always have offers of jobs from other good places. Finally, when they do go, they go to an independent position like your own and become a major competitor! They frequently take away the best projects of your lab, which they of course have a stake in since they were good enough to recognise the good projects. If you are really unlucky, your one-time minion will become much better than you are and will end up extremely famous and be showered with honours for doing what may even be the projects he took away from your lab! Truly, to be eaten by your children is not a fate confined to the Greek myths.

Of course, a good postdoc is infinitely preferable to a bad one. The depths of badness are unbelievable, and it is quite impossible to understand how some people got through their Ph.D.s. Bad people have little idea what the lab is doing, they cannot understand the techniques they are supposed to be using, and have no ability to tell if something has gone wrong. They use large amounts of expensive materials and consume large amounts of your valuable time as you try to explain what is going on and continually simplify their projects to make them more attainable. Because bad people

could easily consume all your time in supervision, they are a serious lia-
bility, being significantly worse than nobody at all. Avoiding the recruitment of a bad person is the highest duty of the lab head, but an extraordinary difficult one to fulfill.

How bad people can be is illustrated by the unfortunate story of my
postdoc Mehmet Hussein. He was of Iraqi extraction but had been educated in Canada and the United States. He wrote to me saying that he was
very interested in working on embryonic induction and could he come to
my lab as a postdoc. Such letters are commonplace, and the normal response in the case of someone who is too far away to interview is to write
back and say, "Fine, you can come so long as you get your own money."
Considering how much is spent on molecular biology consumables in my
line of work, it is curious that there is no easily accessible pot of money to
pay for travel costs for people like this to visit the putative host institution.
Usually, money can be found for interviews if there is "a job" to be filled.
But administrators do not understand the concept of a job that might be
created if the person to be interviewed looks good, so they don't come up
with any travel money.

Of course, there is wide recognition that it is a risk to accept into your
lab someone who you have never met. This is why the formula of "find
your own money" is favoured, because there is stiff competition for personal funding, and it is reckoned that anyone who can get the money is
likely to perform well, or at least be good enough not to cause serious trouble. I thought Mehmet looked okay because he had written a very good
mock grant application in my subject area, which was rather different from
his own Ph.D. topic. His references, when they came, also seemed fine. I
later realised that I had forgotten the first rule of reading American references. British references tend to be reasonably candid, but American ones
are invariably very flattering. The first rule of American references states
that if the referee does not include the phrase, "This person is the best
graduate student to pass through my institution in the last 10 years," then
the candidate is, without any question, *completely useless.*

Mehmet was crafty. He did agree to find his own money, but because of
his Iraqi origins, he was not eligible for most of the personal fellowships
normally open to Americans. In fact, he was only eligible for one. So he
asked me whether there might be a fellowship available at my end, just as
a guarantee in the remote possibility that something went wrong with the
application. At the time I was in the SDS unit in Oxford, and the few SDS
fellowships were allocated by a ballot of the lab heads. But I thought I was
fairly safe because, as often happened, there was a financial crisis going on
and I imagined that there would be no fellowships available. Unfortunately, I was wrong and it happened that there were two available. I duly

entered Mehmet's name for the competition, and in the next ballot he came fourth out of four candidates. "That's okay," I thought. "I don't need to underwrite his application." But then it turned out that the top two candidates just needed half a fellowship each because they were for short extensions of existing positions. Then candidate number three decided not to come to the unit but to go instead to one of my colleagues in the university (he thought he was doing me a favour!). So purely by accident Mehmet had his guarantee. If he failed to get his external fellowship, he would get an SDS one.

Guarantees are tricky things. Although they look as though they cost less than full-scale commitments, you need to budget for the worst case in which they will be taken up. Mehmet did fail to get his external fellowship, so this guarantee was taken up, and he duly arrived with an SDS fellowship. I sensed that something was peculiar as soon as I met him. His communication skills were nonexistent. Mine aren't that great, so for me to perceive a problem in someone else means that they are well on the wrong side of pathological. Foreigners are often quite quiet because of language problems, but this was someone who had been educated in Canada and America, so his English should be perfect. (Perhaps I should say perfectly *fluent,* as we British can never really accept that American English is anything other than a somewhat corrupt dialect.) Actually, there was nothing wrong with Mehmet's English, but he had real problems relating to people in any meaningful way. The only conversation I can remember having with him was when he was expounding about his identification with the wolves in the forests of the Canadian arctic.

It rapidly became clear that Mehmet was unable to do anything correctly. He would make the most elementary mistakes carrying out the simplest procedures. He showed no understanding of what he was supposed to be doing and no understanding of broader scientific principles. To make things worse, he did not ask for advice. He was remarkably arrogant and rude to the other lab members apart from myself. He was particularly unpleasant to my Korean Ph.D. student, who was himself a very courteous and hard-working young man. I gritted my teeth and decided that I would have to put up with him for the two-year duration of the fellowship. In those days I held the mistaken though not uncommon belief that it was not possible to sack someone on a fixed-term contract, so we should just have to tolerate him for the duration. Things really came to a head while I was away in America for two weeks. I remember going into the lab after the journey back, having not slept for 24 hours. "I'm not here," I said as I entered the room. "I'm just collecting my mail." The tension was palpable, and even though I wasn't officially there, I could feel the electricity charging up the entire room. The next day everyone except Mehmet was bang-

ing on my door and insisting that I do something about the situation. It
seemed that all sorts of awful things had happened while I had been away.
They centered on Nancy, my graduate student. She was a very personable
and attractive young woman, but Mehmet had been pestering her for some
months. She gently repelled the advances in the way that attractive women
must, but he totally failed to read the signs, as those with social-skills failure must. Eventually, things had got out of hand, and one day while they
were together in the aquarium, he made a grab for her. Unfortunately, she
had a frog in her hands at the time. She screamed and ran off, the fate of
the frog being unrecorded.

Everyone at work in the 1990s knows that it may be very hard to get rid
of somebody for poor work or for offensive conduct to others, but if there
is the slightest whiff of sexual harassment, there is no problem. The personnel manager for the SDS came down and did the job. He was very
clever because Mehmet could not actually be sacked without receiving a
written warning and then transgressing once again. But he was led to believe that things would be much better for him if he just resigned, which
he did. In fact, he was given three months' notice and almost broke the
unit's budget with his phone, fax, and mail bills over that time because of
all the job applications he was producing. I would not recommend him for
anything, as he was both technically incompetent and useless with people.
So the only reference I would write said baldly that he had been in the lab
for the stated period and had familiarised himself with certain techniques.
I was sure that this would convey the correct message, particularly in the
United States, that he was worse than useless. But amazing to relate, he did
eventually get another job, and in the United States! I have never dared en-
quire how he got on.

The happy mean is represented by the postdocs who are well motivated
towards the research programme, who contribute some ideas of their own
and can get experiments to work, but who are not quite at the lofty heights
of "goodness" such as would cause problems. They will often do the project you want done, do it efficiently, and get it finished and written up.
They will stay for the full three years, and when they have finished, they
will find a respectable job but not become a major competitor. Long and
hard are such paragons sought.

The Student

I was allowed to "sink or swim" with my own Ph.D. project because that
was in the distant years of cornucopia in the early 1970s. But in the serious 1990s, when accountability is everything, Ph.D. students have to have

projects that are guaranteed to succeed; otherwise, it is a black mark against the supervisor for wasting resources.

Lab heads must be very careful about selecting projects, because they must fulfill certain requirements:

1. The project must fit into the lab's existing programme, so the student can benefit from the surrounding expertise and can eventually contribute something to the lab's work.
2. On the other hand, it must be relatively self-contained, so the student is protected during her three years from predatory postdocs in the same lab who want to clean up a topic and publish it fast in a good journal.
3. The project must not be too difficult or risky because the student must be protected against the risk of total failure. Total failure nowadays usually results from the failure to isolate or identify some clone on which the rest of the project depends.
4. It can't be absolutely in the front line of competitive science, because aggressive people in other labs will clean up the topic long before the student's three years are up.
5. On the other hand, it must be interesting and important enough to engage her attention and impel her to put in the evenings and weekends expected of junior scientists who want to get on.
6. It should involve a reasonable range of techniques so that the student gets some breadth of technical training.
7. It needs to be both long enough and short enough to fit into the allotted three years and also have enough coherence as a single project so that it can be written up in the form of a thesis after this time.

These constraints are considerable and do very much limit the utility of research students in a front-line competitive programme. In addition, of course, graduate students are beginners at research and usually don't have anything ready to publish until the end of their three years. Most of these problems do not apply to postdocs, which is why they make up the basic cannon fodder in most "good" labs, and research students in such places are often in a minority.

The Technician

When I entered science, there was a species of individual called the laboratory technician. But in many places this species has either totally disappeared or is becoming dangerously rare. Has some great environmental disaster occurred to bring this about? Is it a massive attrition due to radiation, chemical poisoning, or deadly recombinant biohazards? In fact, the extinction has arisen from the quite different, but just as deadly, process of grade inflation.

The idea behind the job of technician is that there are many routine jobs to do around the lab: feeding animals, making up stock solutions, performing tests or assays of a standardised character, keeping up catalogues of chemicals or of clones and antibodies made in the lab, and so on. If scientists do these things, they have less time for such indisputably creative tasks as thinking of new experiments, performing difficult manipulations, inventing new pieces of apparatus, or thinking of new explanations for natural phenomena. (I dissemble; most scientists actually spend most of their time applying for grants, preparing teaching materials, supervising other staff, or going to committee meetings. They are lucky if they have time to think of a new experiment in the car on the way home.)

So, up to about the 1960s, a lab might contain some scientists, who would be Ph.D. graduates, or in industry quite often just first-degree graduates, and some technicians, who would normally have started as school-leavers and have been trained on the job to do a range of supporting tasks. The growth of mass higher education, even if it was rather belated in Britain, gradually reduced the pool of competent school-leavers by converting them all into university graduates. So technical posts started to be filled by graduates, and it became difficult to find nongraduates at all. The individuals concerned were no better qualified to do the jobs that needed to be done, but they had a better all-round scientific education that enabled them readily to perceive that scientists had high status while technicians had low status. Because they themselves were graduates, they reasoned that they should have high status as well.

Gradually, the academic institutions started to change job descriptions; there were fewer and fewer jobs for technicians, but more and more for "research assistants" or "research officers." As time went on and higher education expanded still further, the workforce became even more highly qualified, and Ph.D. graduates started applying for the technical posts. Ph.D. graduates usually don't want to "waste" their Ph.D.s (i.e., merely settle for the use of the title of "Dr." when writing to the bank manager); they want jobs in science. And since there are always very few academic jobs, they apply, now in large numbers, for the jobs as research assistants and research officers. At least they have had three years of practical experience working in a lab, so unlike B.Sc. graduates, they can usually make up solutions without mishap, but needless to say, the vast bulk of necessary laboratory technical work is now beneath them. They want to do a "project," not to feed the rabbits.

To some extent this difficulty can be overcome by relabelling the necessary routine tasks and building into them some small element of controlled variation so that they become "research projects." But on the whole it is a real nuisance. Meanwhile, in the most advanced institutions, such as the Serious Disease Society, titles such as "research officer" or "research assistant" have now gone the way of "technician." They have lost status and are seen as something rather demeaning. Indeed, in the SDS, all the former "research officers" are now "scientific officers." Many of these posts are held by Ph.D. graduates, and all are convinced that without a Ph.D. they will have little chance of promotion. So laboratories of this type now have all chiefs and no Indians. The process of genocide that destroyed the laboratory technicians and is now removing the research officers will doubtless in due course come for the rest of us, until finally everyone in the organization is "Director of Research." I think Gilbert and Sullivan has something to say about this. In their operetta *The Gondoliers,* the two kings with their staunchly republican principles had made everyone into a noble or courtier. But "when everybody's somebody, then nobody is anybody." Meanwhile, the routine jobs still need to be done. In fact, as science becomes more complex, there tends to be more and more routine technical work. So who does it? On the whole, it is now divided among the Ph.D. students and the postdocs. Never mind, it will be good training if they ever need a job as a scientific officer.

Tenure

Tenure has tended to be more of an issue in the United States than in Britain. This is because in the United States a faculty member will not be made permanent until several years into his appointment, usually in his or her late 30s. In Britain university staff mostly have tenure, while full-time researchers on personal fellowships do no have it and cannot get it without taking a university job. However, there are some institutions that do give tenure to full-time research staff. These are the Government's Research Councils that run various institutes, and also some medical charities such as the SDS. In these cases the tenure system closely resembles the American model. "Tenure track" appointments are made for five to six years, after which the researcher is assessed for tenure, and if he does not get it, he is out.

Tenure used to have a special legal significance meaning that you could not be sacked for anything except "flagrant and persistent immorality." The object of this special status was to protect academics who said or wrote things inconvenient for their political masters. In modern Western society, there is very little political interference with what academics say, but there is a great deal of government financial regulation of their institutions. In the United Kingdom this has led to the abolition of tenure and its replacement by a formula such that you can be sacked for failing to turn up for work, refusing to carry out reasonable duties, or if your institution goes broke and has to make you redundant. Such changes are also percolating through the United States in a more piecemeal fashion. In other words, tenure now means just an ordinary employment contract such as is found in most of Western society. However, even in the hungry 1990s, it is still unusual for people to be made redundant from universities and institutes, so these jobs are more secure than most.

Why have tenure in the first place? Many department chairs would be happy if all their lab heads were on fixed-term contracts, as they could then get rid of the ones they didn't like and replace them with more amenable individuals. Because research scientists have to work independently, it is difficult to specify anyone's duties; so once they have tenure, there is little chance of stopping someone from lapsing into indolence. However, it is necessary to have a core of permanent senior staff to provide some continuity and management. It can also be difficult to recruit people at senior levels if you cannot offer tenure. Young people will put up with short-term contracts, but if you want to recruit someone at senior level, they may well have tenure already and be unwilling to give it up.

To get tenure in an American university or a U.K. research institute, you have to apply for it, which involves having a very good CV, a publication list including lots of papers in fashion journals, and writing a very good proposal about your future work, which is sent to several external reviewers. You then give a lecture to the appointment committee, and they grill you for several hours to see how good you are before making a decision. If you don't get tenure, then you have to look for another job, usually at a lower-status institution than the one you are about to leave. Tenure proceedings are naturally riddled with small-p politics, personal antagonism, and jealousy.

I remember a salutary tale of two tenure-track individuals called George and Ben. Both worked in a laboratory at the SDS Richmond Hill Institute, and both were good scientists. But when it came to tenure, one of them did everything right and the other did everything wrong. They shared the same basic and serious problem, namely the *slime moulds* that were the research topic of their laboratory. Slime moulds are peculiar creatures that live in the leaf mould on forest floors. Their life cycle consists of a protracted stage in which they exist as single-celled amoebae, and a short fruiting process that they undergo if they run out of food. In the fruiting process, the amoebae aggregate to form a multicellular blob called a slug, which crawls around for a bit then turns into a fruiting body. This releases spores that can, in their turn, later hatch into new amoebae. There are many interesting problems raised by these harmless little creatures: how the cells aggregate, how different parts of the slug become the stalk and spores of the fruiting body, and how the slug can still form a normal fruiting body even after large parts have been cut away.

George and Ben both started out in a slime mould lab run by Jack Large, the only permanent appointment eventually made by our director after he had cleaned out the Richmond Hill Institute (see Chapter 2). Large was tall, with curly fair hair and a rather absent-minded manner. But he had a keen intellect, and his slight Welsh accent would come to the fore when he was making a decisive point. The slime moulds were much loved at Richmond Hill, but they were also a problem: because they do not cause any serious diseases, they were seen as rather irrelevant by the central directorate of the SDS, who would have much preferred to be rid of them. When our local director himself departed for a job in the United States, he left his protégés exposed, and none more so than George and Ben, who had both unwisely delayed applying for tenure until after he was gone. Jack Large already had tenure and so could not easily be removed. But he did manage to contribute in no small way to the difficulties for George and Ben, because over the years he had become rather involved with the religious cult of the Bhagwan Shree Rajneesh, and this involved spending long periods of time in India sitting at the feet of the master.

So, what with the unattractiveness of slime moulds to the central powers, and the embarrassment of a senior scientist who spent much of his time in India serving the Bhagwan, the tenure prospects of George and Ben looked pretty bleak. But George knew just what to do. As soon as it became clear that the American superstar James Samson was never going to show up, George moved himself into part of the space that had been reserved for him and took with him the others in Large's lab associated with his projects. This had two benefits. First, he was now on a different floor of the building from Large and thus somewhat removed from the cloying embrace of the Bhagwan. Second, he now had a spacious lab of his own, so it looked as though he already had tenure. It is, of course, always easier to get tenure if everyone thinks you have it already. Then George started up a new line of work on *Xenopus,* which was less unacceptable to the central directorate than slime moulds. *Xenopus* also doesn't cause any serious diseases, but it is at least a vertebrate and so shares most of the same genes, cell types, and developmental mechanisms with humans. Finally, and most important of all, George made himself into the SDS's master of cDNA cloning. This was very important indeed. Although the advent of cloning in 1973 had made it possible in principle to amplify any desired gene in any organism to a useful quantity and to study the properties of the protein product, the techniques in the early AC era remained extremely tricky, and people could easily spend years of fruitless work without isolating the clone that they wanted. Because George was a master of this important technology, he always had a long queue of people outside his office wanting advice about their problems. They came not just from Richmond Hill but also from the SDS Central Laboratories and just about everywhere else. Everybody knew that George was a vital resource for the SDS and just had to have tenure. When George finally put in his tenure application, his future research plans consisted almost entirely of work on *Xenopus* with a small footnote about finishing off one or two little things with the slime moulds. Sure enough, he did get tenure, and within months the frogs had gone, the people working on the frogs had gone, and the entire lab was given over to the unloved but exceedingly grateful slime moulds.

What about Ben? Ben failed to study George's progress closely enough and ended up doing everything wrong. He stayed in Large's space and was closely associated with Large and therefore, to external eyes, also with the taint of the Bhagwan. His work had also gone well, and he had isolated a novel substance that controlled whether the cells of the slug differentiated into stalk or spore. But his tenure application, promising many more years of fruitful work on the slime moulds, was just too much for the SDS and was rejected. Fortunately for him, it was understood elsewhere that the SDS directors had developed a phobia about slime moulds, and he was able to move to another institute in Cambridge, which unusual for a tenure victim, had a higher status than the institute he had left.

Shortly after these events, Jack Large himself was persuaded to resign. He went off to join the Bhagwan for a while but after a couple of years was obviously missing the excitement of science and came back as an externally funded research fellow in a university. This meant that the slime moulds kept their toe-hold in the SDS only through George's lab. George stayed for 10 long years, spending large amounts of the SDS's money on his slime mould research, which was undoubtedly of world-beating class for those who were interested in that kind of thing. Eventually, the wanderlust of a midlife crisis dislodged him and he left for a professorship at the University of Portree. At long last the SDS was free of the sticky slime moulds, and the central directorate could breathe a sigh of relief. Needless to say, George's successor worked on something much closer to serious diseases.

Laboratory Life and Death

Biological scientists work daily in a fog of radioactivity, toxic and carcinogenic substances, inflammable solvents, and genetically engineered microorganisms. You might think that violent death would be an everyday occurrence and that our average life expectancy would be somewhat lower that that on the western front during World War I. But surprisingly, accidents are very rare, and serious accidents almost never happen. The reasons for this miraculous survival pattern can only be explained by applying to our activities the new science of "relative risk assessment." The basic principle of this new science may be summarised quite simply by saying that, compared to smoking, not only the biological laboratory, but all other human activities look fairly healthy.

Radioactivity

The gulf between public perception of risk and real risk is nowhere greater than it is on the subject of radioactivity. Radioisotopes were introduced into biochemistry when they first became available in the 1940s. Initially, they were used mainly as *tracers*. A normal substance that had been synthesized to include one or more radioactive atoms would be fed to an animal, a plant, or a flask of bacteria. After a while the organisms would be ground up and analysed to find which new substances had become radioactive. These must have been formed from the tracer substance by metabolism, and so by using many different labelled precursors and making many measurements at different times, it was possible to work out the whole complex map showing the synthesis and breakdown of substances by living organisms. Nowadays, in modern molecular biology, isotopes are used mainly to visualize small amounts of material because the means of detection of radioactivity are much more sensitive than chemical methods. For example, using radioactivity it is possible to "see" a single gene in a sample of total DNA extracted from cells, using a method called the "Southern blot."

Because radioactivity is very easy to detect, radioactive contamination is also easy to detect. This is done using small portable Geiger counters, or monitors, that emit a crackle proportionate to the amount of radioactivity detected. It is good practice to monitor the area in which you have been working to check that there are no spots of contamination left behind. But in every laboratory, there is a mysterious individual called the "phantom contaminator." He comes in when there is nobody else around and deliberately puts spots of radioactivity in various places carefully chosen to produce the maximum embarrassment. Everyone who has been working in the area in question will vehemently deny having caused any contamination, so it is clear that the phantom contaminator must be to blame. But somehow, like Macavity the mystery cat, he always seems to evade detection.

In the SDS's Richmond Hill Institute, each year an unfortunate technician (or as they are now known, scientific officer) became the Radiological Protection Officer (RPO) and had the job of filling in the innumerable forms to do with storage limits, disposal limits, individual records, film badges, and so on, as well as trying to enforce good practice on the workforce and chasing the phantom contaminator. I remember a particularly effective RPO called Sylvia. She was a tall woman with blonde hair and an outgoing but forceful personality. Until she became RPO, she was very popular, being one of those people who never stops talking and who can be heard all the way down the corridor leading to the coffee room. But she was unlucky enough to experience a visit from the Health and Safety Executive (HSE), a government department dealing with occupational safety, shortly after the phantom contaminator had visited the hot room. I should perhaps explain that the hot room is, like the cold room, a feature of every biological laboratory. However, whereas the cold room is really cold, about 4 degrees centigrade, the hot room is only figuratively hot. Its "hotness" lies in the levels of radioactivity allowed to be handled there, which are greater than elsewhere. Indeed, most vials coming from the radiochemical suppliers contain more than the levels permitted for ordinary labs, so they have to be opened and the contents dispensed in the hot room. Taking into account the general clutter in the hot room, the droppings of the contaminator, and the generally dodgy state of the paperwork, the HSE men were rather hard on Sylvia and hinted that standards were so bad that, if this sort of thing continued, permission to work with radioactivity might be withdrawn from the institute.

After this incident Sylvia introduced a reign of terror. She insisted that the fate of every radioactive atom had to be accounted for on paper. So the limiting factor in performing an experiment now became the number of forms that had to be filled in each time a vial of radioisotope was approached. But she was right. She had learned the hard way the cardinal principle of safety management. It doesn't matter how many corpses are piled up on the floor; the important thing is to have the *right documentation* ready when the inspectors come.

The main problem with radioactivity is, of course, its disposal after the experiments are over. Such is the pathological fear of radiation that members of the public, normally so reluctant to pay taxes, will pay billions to ensure that not one subatomic particle comes anywhere near them. The main problem in this regard comes from solid rather than liquid waste. Because most biochemistry is done in solution, the amount of radioactivity in the liquid waste is usually much higher. But this does not matter because the allowable limits for disposal of radioactive liquids down the drains are also high, presumably because the radioactivity is quickly diluted down to undetectability by the rest of the sewage. So the public remains blithely unaware that most radioactive waste simply gets tipped down the sink and that this process is perfectly legal. Solid waste is usually much less radioactive since it consists mainly of gloves and paper tissues that have just been *near* a radioactive material. But because it cannot easily be disguised, it is regarded as totally deadly by the public and has to be buried in a special vault guaranteed to last hundreds of millions of years. This is very expensive compared to washing it down the sink, so Sylvia installed a machine called a macerator that would eat all manner of glass and plastic and reduce it to the consistency of soup, which could then, perfectly legally, be flushed down the sink. She was often to be seen bending over its gaping maw watching anxiously as it ground and crunched away at its food. The macerator earned its keep in no time and was very useful in reducing the growth rate of the solid waste stockpile. But still the unmaceratable solid waste kept piling up.

The solid waste problem really came to a head with the closing down of the local low-level disposal dump during the time that we were in Oxford. Some 20 or 30 years previously, it had been possible to open dumps for innocuous low-level waste in specially designated sites around the country. Because it was just taken away and dumped, nobody gave much thought to where their radioactive rubbish was going, until suddenly one day the local authority announced that their own low-level dump would be full in a few weeks. By this time the idea of a *new* radioactive waste dump anywhere outside the perimeter of a nuclear reprocessing plant was enough to create a major constitutional crisis. The University Safety Officer then caused havoc by recommending the "dustbin" method of disposal. It turns out that according to U.K. law, you are permitted to dispose of one millicurie of radioactivity per cubic metre of ordinary domestic rubbish (this works out to about two radioactive atom disintegrations per second per cubic millimetre). Sylvia called a special meeting to explain the new dustbin method. Most of the foreign postdocs present at the meeting were amazed and complained loudly, considering that the British method of dealing with such things had declined to the standards prevalent in the former Soviet bloc. However, it turned out that none of them actually knew what happened to low-level waste in their own countries.

We wanted to store the waste ourselves for a while, because ^{32}P-phosphorus and ^{125}I-iodine, which are the most dangerous isotopes in common use for molecular biology, have half lives of only a few weeks, and storage for a year would effectively mean that the waste had all decayed. Unfortunately, Sylvia had already looked into this and found that storage is strictly forbidden unless you have specially designated lead lined chambers, negative pressure, multiple alarm systems, 24-hour armed guards, and so on. Actually, the dustbin era did not last long. Not surprisingly, various people around the university informed the press, and we soon read gruesome newspaper headlines implying that the university was chucking lumps of plutonium into the domestic rubbish every day. Finally, the inevitable happened and arrangements were made with the supplying companies for the waste to be collected and, at great expense, transported to some special vault at the nuclear reprocessing plant at Sellafield, which, we are reliably guaranteed by the experts, will suffer only one serious accident every 10^9 years.

Something that people find very hard to understand about radiation is that it occurs naturally and is everywhere all the time. The heavy elements that are naturally radioactive are widely dispersed in rocks, soil, or clay, and whenever you switch on a monitor, you will hear the clicking sound indicating the presence of background radiation. Considering that we are exposed to this background level 24 hours a day for a whole lifetime, short exposures to the quantities used in biological experiments make very little difference to the overall dose. In fact, on the day that the Chernobyl cloud was supposed to have come over, I took home a monitor and found that the brick walls inside my house were more radioactive than the grass outside.

One day, I got into trouble with Sylvia because a small amount of ^{32}P had been detected on my dissection table, which was made of solid slate. I tried to blame the phantom contaminator, but she did not believe me. She was very annoyed because the contamination set off the siren on her portable monitor, which emitted a tone specially selected by psychologists to cause maximum alarm and anxiety. I scrubbed away at the spot and eventually maintained that the contamination was no longer significant. It did indeed still set off the alarm, but by this time I had worked out that the heavy slate top of the table itself emitted enough background radiation to take the monitor to 75 percent of its alarm threshold and the residual contamination was supplying the other 25 percent.

Finally, there was the day when we thought our own Chernobyl had arrived. Someone from another lab borrowed our scintillation monitor, which is a type designed for the detection of gamma ray emitters such ^{125}I. He came back a few minutes later asking why it gave such a high reading when pointed in the general direction of the windows. The windows! But that's outside; there is no radioactivity there (except for the ubiquitous cosmic rays nobody cares about). We wondered if there really was radioactivity outside. A small posse went up onto the roof and pointed the monitor to the sky. There was a deafening crackle. Christ! What is it! we all thought. We pointed the monitor downwards towards the pools of rainwater lying on the roof—only normal background. So there was radioactivity in the air but not on the ground. This meant that a radioactive cloud had just arrived. Was it from a French nuclear station that had just exploded? Or maybe it was some disaster in another university department that people were keeping quiet about. We decided to check in a different place so went down to the front door. Sure enough, the crackle was there too, where conference delegates were calmly going to and fro, blissfully unaware of the deadly radiation on all sides. We decided it was unhealthy to stay outside and retreated back into the building. What should we do? I decided the first step was to ring the University Safety Office. After all, the cloud could be local, and maybe nobody actually knew about its release. Of course, the radiological safety officer was "in a meeting," so I left an urgent message with his secretary about a deadly radioactive cloud drifting over the science area. But as soon as I had done this, the mystery was solved. Scintillation monitors, as their name suggests, are not direct detectors of radiation like Geiger counters. Instead, they detect light, which is emitted from a fluorescent substance when it is impacted by gamma rays. Our monitor had a tiny hole in its tube, which meant that a small amount of daylight could enter. When it was pointed to the sky, enough light got in to give a good crackle, but when it was pointed to the ground, the light penetration was not enough to give a signal. We all breathed a great sigh of relief, and

I rang the Safety Office again to eat humble pie about the deadly radio-active cloud.

Biohazards

In the year 0 of molecular biology, when the world was young and cloning was invented, terrible scenarios gripped the minds of some of the scientists involved. "Supposing we cloned a gene that caused cancer, and put it into *E. coli* (the usual bacterium used for propagating clones), and it got out of the lab and infected people's guts. Then what??!" Doubtless, many of those in the know had recently seen the film of Michael Crichton's *The Andromeda Strain,* a story that depicted the terrible consequences of a plague arriving from space. Perhaps genetic engineering could unwittingly create new plagues by rearranging genes from different species in wholly new ways! In July 1974 Paul Berg and 10 other molecular biologists engaged in the work wrote letters to the journals *Science* and *Nature* calling for a moratorium, or pause in further experimentation, until proper safety procedures could be devised.

A special conference was held in Asilomar, California, in February 1975 to discuss the risks and formulate some guidelines for resuming the work. I remember some of my colleagues from Edinburgh going excitedly to this meeting. When the dust had settled, it was clear that any lab that wanted to do molecular biology was in for serious expenditure. There were four categories of "physical containment" prescribed for different classes of manipulation. To do anything significant, like cloning new genes, you needed a "P3 facility." This was a sealed room with an antechamber for changing and washing. The air was under negative pressure and filtered on extraction. Everything that came out of the lab (except the people) had to be autoclaved in order to sterilize it. Being keen to get on with the new technology, the SDS built a P3 lab at our institute. It just happened to be adjacent to my lab, and there also happened to be a fire escape door leading directly into my lab that, as it was always closed, had none of the security measures applied to the main entrance. This fire escape door contained a window, so you could clearly see into the mysterious P3 room with its beneficent ultraviolet light on the lookout for any Doomsday bugs. But the presence of the fire escape did pose us with a potential moral problem. Suppose there was, in fact, a biological disaster? Suppose we saw our colleagues, their faces half eaten away by green fungus, scrabbling at this door and trying to get out into our lab? Should we let them out and start a global pandemic, or should we harden our hearts, leave them trapped, and press the incineration button? A tricky dilemma, with which, fortunately, we were never faced.

In some ways the beginning of genetic engineering was a model of care and responsibility in the introduction of a new technology. When it started, very few people could really understand the techniques or the implications. To slow things down and proceed with extreme caution for a few years really was the right thing to do. Unfortunately, the effect of all this was to persuade the general public that genetic engineering was extremely dangerous, and it soon became bracketed in the popular mind with that other deadly killer, radioactivity. It is interesting to step back and look at the actual risks. In one year in the United Kingdom approximately 40,000 people will die of smoking-related diseases. About 4,000 will die in road accidents. But apart from the consequences of medical radiotherapy, nobody at all will die from exposure to radiation or to genetically modified organisms. Perception of risk by the general public clearly has a long way to go.

However, in one regard the scepticism of the general public about official risk assessments is fully justified, and this is when officialdom attempts to calculate numerical risks. I learned about all this when I did my stint as Biological Safety Officer at our institute. By that time the fuss had died down and all the types of experiment had been reclassified to a degree that made it possible to do some work. We dealt with a body called the Genetic Manipulation Advisory Group (GMAG), which later became absorbed into the Health and Safety Executive (HSE). During the initial period, there had been the most absurd anomalies because P3 containment had been required for cloning of harmless pieces of DNA, whereas genuinely deadly microorganisms were being routinely handled in hospital pathology labs with facilities approximating to P1 or P2, representing much lower levels of containment. At the stage when I had to administer the system, each type of manipulation was given a "risk factor," which attempted to calculate the probability of "serious biological damage" to humans. It had been decided that DNA itself was not dangerous; only the proteins coded by the DNA might be. So the overall risk factor was obtained by multiplying together three numbers representing the estimated risks for access, expression, and damage:

Access refers to the likelihood that the bacteria used to propagate the clones could infect humans.

Expression refers to the likelihood that the plasmid, or self-replicating piece of DNA containing the gene sequence in question, could permit synthesis of the protein coded for by the cloned gene.

Damage refers to how dangerous the protein itself might be.

A typical experiment done in developmental biology involves the use of enfeebled bacteria, which are unable to transfer genes to other bacteria or to grow outside the lab (risk factor 10^{-6}). If the plasmid is not designed for protein expression, some very unlikely recombination events would have to occur to make this possible (risk factor 10^{-6}). If we are dealing with a typical genome, the chance of a random piece of DNA being biologically active as a hormone or growth factor is quite low (risk factor 10^{-6}). Readers will therefore see that random cloning of bits of DNA from a mammalian source in a system not designed for protein expression would carry the following overall risk:

$$\text{access} \times \text{expression} \times \text{damage}$$
$$10^{-6} \times 10^{-6} \times 10^{-6} = 10^{-18}$$

This is what we wanted. During my reign experiments with a risk factor of 10^{-18} or less could be conducted under "good microbiological practice." I did sometimes wonder whether the sloppy and chaotic conditions that usually exist in molecular biology labs did not correspond better to a description of "bad microbiological practice" and require a risk factor more like 10^{-21}. However, mine was not to reason why; I just had to keep the paperwork in order, which meant keeping a list of registered workers, sending in annual returns of the types of experiment performed, and making sure that all the risk factors came to 10^{-18} or less.

But what, I sometimes wondered, did this number of 10^{-18} actually mean? Was it one serious biological disaster every 10^{18} years, or one disaster every 10^{18} experiments, or was there one deadly clone in every tube containing 10^{18} different clones? The written guidelines were quite precise on how to calculate the denominator in this factor but curiously vague about the numerator. Of course, real risk assessment always depends on understanding. Today, now that the regulations have been relaxed still further, we often grow bacterial cultures specially engineered for maximal production of growth factors. We did, and do, treat the cultures, and the expression clones themselves, with some prudence, and as far as we know, nobody in our lab, or elsewhere in the world, has suffered any ill effects.

There are genuine risks associated with work in laboratories, just as there are with any human activities. But it is likely that the real risks are highest for the rather old-fashioned and dull things like inflammable solvents, electric shock, chemical toxicity, and broken-glass injuries. The best prevention, in these areas as in the more exotic, lies in making sure that the workers understand what they are dealing with and where any potential hazards are. Effective education is more likely to prevent disasters than any number of correctly filled in forms.

Animals

No biological or medical research would be possible without the use of animals. This is an unpalatable truth to many who would like to believe that there are "alternatives" that are as good or better. Alas, there are not. Scientists want their experiments to work, and if there exists an effective method that works, and it is legal, then they will use it. Tissue culture cells are certainly very useful for some purposes. For example, as we saw in Chapters 3 and 5, they can often be superior to animal tissues as sources of growth factors. They are also much easier to use than whole animals for a wide variety of cell biology and biochemical work. But there are things whole animals do that tissue culture cells do not do, such as have organs composed of many cell types working together, or show particular types of behaviour, or get diseases. If one of these sorts of thing is under investigation, there is really no alternative to the use of animals. Furthermore, it is often forgotten that the tissue culture cells are themselves obtained from animals and that they almost always need animal products, such as serum, in order to grow in culture. Another "alternative" that we are frequently offered by the "animal rights" activists is computer simulation. Computer simulation can be very useful, particularly if you want to know whether you can reproduce the behaviour of a complex system by modelling it in terms of a number of parts, each of which is individually understood. But you need data to put into the simulation, and that data usually needs to be collected, directly or indirectly, from animal experiments.

Much of the public debate about animal experimentation is conducted in terms suggesting that it is all about the safety testing of new drugs. To be sure, new drugs need to be tested, and most of these tests need to be done on animals. But there is much more to the development of a new drug than safety testing. In fact, the first and most important step is one that is often ignored altogether, namely, the discovery of the basic biochemical or physiological processes with which any new drug will have to interact. This is quite likely to have been done years before attempts to develop relevant drugs are started, and probably in a university lab rather than in a pharmaceutical company. If you look at the development of any of the major classes of drug in use today—from antidiuretics to anti-inflammatories to antihypertensives—you will find that a considerable amount of animal experimentation was essential to discover the basic physiology and to identify the therapeutic targets of the drugs, long before the drugs themselves were synthesized or the development with its associated safety testing was commenced.

The public image of animal experimentation is a rather nineteenth-century conception of experimental surgery, with the animal strapped down, probably unanaesthetized, and the experimenter covered in blood, cutting away with grim determination. Fortunately, it is not like this at all. In the United Kingdom animal experimentation is exceedingly tightly regulated, and in the United States the state or local regulations now also tend to be quite strict. If you are a farmer, you are allowed to chop the tails or testicles off your livestock with no regulation at all. If you are an angler, you are allowed to impale your fish repeatedly with rusty hooks. If you are a domestic cat, nobody will turn a hair if you rip apart mice or birds, playing with them as the disembowelment proceeds. However, if you happen to be a scientist, you will not be allowed to so much as change the diet of a laboratory mouse without filling in a stack of forms several inches thick. I know about all this because in the United Kingdom it is necessary to have a full set of licences in order to inject frogs with hormone in order to get them to lay eggs. The institution needs to be licenced, the rooms in which injections are permitted need to be licenced, the overall scientific project needs to be licenced, and every individual who injects the frogs needs to be licenced. Furthermore, every injection must be carefully recorded because it counts as one animal experiment. So if one frog is injected three times in a year, which is about the optimum frequency for good eggs, this counts as three animals used for scientific experiments.

Under such a strict regime, the frogs themselves are naturally well looked after. Of course, it is also in our interest to look after them well because we want good eggs. So they live in glass tanks where they are clearly visible and the least sign of disease can easily be recognised and attended to by their named veterinarian. They get four litres of water each, which is changed three times per week. Temperature and water quality are closely monitored. They get fed twice a week, and if this doesn't seem very often, remember that they are cold-blooded and don't take a lot of exercise. In fact, they get too fat if they are fed more than twice a week. Under such luxurious conditions, individual frogs may last as long as 10 years. It is certainly a far cry from life in the wild. *Xenopus* come from South Africa and often live in small pools of extremely dirty and foul-smelling water on the edges of cattle fields. Nobody there gives them any food, so they have to catch whatever worms and insects come their way. When they arrive in our lab, they are often rather thin and may have parasites or infected ulcers. We treat the parasites, give the frogs antibiotics if necessary, and fatten them up for egg production. Obviously, we wouldn't stand much chance of keeping our Home Office animal licences if we kept the frogs in anything like the conditions they endure in the wild.

These considerations are not well known to the so-called animal rights activists who seem to take a delight in planting bombs under scientists' cars or demolishing laboratories with sledgehammers, rather than finding out what luxurious conditions lab animals actually live under. But because of them, all animal facilities have to have very tight security. I remember one embarrassing result of this when I happily went into the mouse house after hours. Once I had opened the door, I realised that I had to punch in a code into the alarm system to deactivate it. I had 60 seconds to remember the code! No bomb disposal officer could have been more frantically punching in half-remembered numbers than I. Alas, I could not get it right, and the alarm went off like an air raid siren. I sheepishly went to the departmental office, confessed what I had done, and asked for the alarm to be turned off. Impossible! There was no way of turning it off, because if there were, the animal rights terrorists could turn it off themselves. Worse, it was connected directly to the police station. After a few minutes, the police swarmed into the building looking for thugs with balaclavas and pickaxes intent on "liberating" the animals. They didn't seem very interested in me when I confessed to having let off the alarm by accident. Then I understood that they must have instructions to ignore anyone like me because I might be there to make them go away and let the terrorists get on with the job. So I had to wait until they had searched the whole building and declared that nothing was amiss. After this experience I lost confidence in my ability to remember secret codes and always seemed to arrange things so that I didn't have to go into the mouse house after hours.

Britain not only has very strict laws about animal experimentation, it is also very difficult to import animals. Importation of mammals is particularly difficult because the British public is terrified of rabies. Britain has been free of rabies for over 100 years, and because it is an island, it is possible to keep it that way by controlling animal movement at airports and seaports. The recently constructed channel tunnel joining England and France has large numbers of death rays and poison-dart guns, in addition to the normal grilles and traps to prevent any wild animals running the 20 miles down the railway track. This is because the British public is convinced that once a single mouse enters the country from abroad, the entire human population will die of rabies. Curiously enough, in those neighbouring countries where rabies does exist, nobody seems to have died of it for many years. Conversely, the British seem quite happy to eat BSE-infested beef and die of Creutzfeldt-Jakob disease, and they can't understand why the filthy foreigners will not buy their beef.

The standard rule for mammals such as domestic pets is that they should be kept in quarantine for six months to ensure that they are rabies free. This effectively prevents anyone taking their animals on holiday with them to the Continent and ensures a stable trade for kennels and catteries that accommodate the pets when necessary. There is no exception for laboratory animals, except that it is permissible for them to be quarantined in a suitably equipped laboratory animal unit. Just because mice are small does not mean that they do not have to serve the full sentence of six months. It is like the story of the 70-year-old man convicted of murder and sentenced to 25 years imprisonment. "I can't stay in prison that long. I shall be dead long before then!" cried the distressed prisoner. The judge was in no mood for mercy and replied: "Then you must serve as much of the sentence as you can." A mouse of six months is quite an old mouse and is not much use for reproduction. As it would be impossible to establish a colony of imported mice using only geriatric survivors of the quarantine period, it is permitted to breed them during the quarantine, and for some curious reason that defies logic, it is not necessary to confine the offspring.

My colleague Dr. Sen wanted to import 12 mice of a particular transgenic constitution from Japan. They came from a special germ-free facility in an immunology institute, but merely being reared entirely under sterile conditions is obviously not sufficient protection against importing rabies into Britain, so the full force of the law descended on the 12 mice. After the mountain of paperwork was completed came the financial considerations. It just so happened that the room designated for quarantine was the largest room in the Animal House and was the one already containing the rest of Dr. Sen's stocks. It was impossible to use another room because only this one was in the right position for all the extra barriers and security devices needed for a quarantine facility. Of course, no other animals may be kept in the quarantine room during the quarantine period; otherwise they too would be classified as rabies carriers and be confined for six months. So Dr. Sen's other mice were moved to a vacant room and the 12 Japanese mice went into the quarantine room. In most places this might be the end of the story, but it just so happens that the university was operating a system of "cost centre accounting" whereby each cost centre pays rent on its space. For facilities like the Animal House, the rent is charged to the user. So Dr. Sen's mice almost broke her financially, as each was occupying and paying for as much space as would normally be occupied by several hundred mice. There is nothing like a stiff financial incentive to produce results, so as soon as the mice were in the door, they were set up to mate, and as soon as the offspring were weaned, off went the unfortunate Japanese imports to the incinerator. They were just too expensive for their own good.

Computers

As mentioned in Chapter 4, the main output of scientists consists of their publications. These are called papers and are published in a great variety of scientific journals ranging from the slick and fashionable (*Nature, Science, Cell*), through the solid and reliable (e.g., *Journal of Cell Biology, Developmental Biology*), to the scrapers-up of any old garbage left over in the notebook (mainly short-life journals launched by commercial publishers to make a quick buck). Up until about 1981, all scientific manuscripts were typed by secretaries. They were usually typed in a first draft, which was circulated for comments and revised, and then in a final version for submission to the journal. It was not unusual for multiple copies to be made using carbon paper, although the use of this traditional material had declined with the general availability of the photocopier.

In 1981 the first word processor arrived at the SDS. Although we didn't have one in our satellite institute, I was writing my book *From Egg to Embryo* at the time. So I sent the handwritten sheets into the centre for typing into this fabulous machine. About a year later, we got our own word processor—in those days a large machine filling up half the main office—and everyone immediately wanted all their manuscripts to go onto it. The great attraction was the unlimited scope for corrections. No longer need one worry if the first draft was a little sloppy; with a word processor, you could have any number of drafts! Because there was only one machine for the building, a long queue developed, and for a while, it was much quicker to get things typed the conventional way. Our two secretaries, Brenda and Julie, had both mastered the arcane mysteries of the word processor but were now deluged with requests to make modifications, because of course everyone knew that with a word processor, you could make as many changes as you liked. Although both women had almost infinite tact and patience, this was really too much, as they were spending all day making small corrections in already-typed manuscripts. Fortunately for them, they were rescued by the age of the personal computer.

The director of the SDS had a great liking for computers, since he was himself a human geneticist for whom doing science meant not so much doing experiments, but rather collecting and analysing vast amounts of data. When he arrived, he not only brought a large computer with him but also set up a department to run computing services for the society. This had one unfortunate consequence because it entrenched in the organisation people with a vested interest in the *upgrade*. When the microcomputer era started, the director of computing happened to be a believer in IBM compatibility. After a couple of years, he was sacked and the new one was a fanatic adherent of the Apple Macintosh computer. Within a very short space of time, everyone had scrapped their perfectly serviceable but now unfashionable IBMs and acquired sleek new grey Macs. There is no doubt that when Macs had a monopoly on the graphics interface, it was easier for a complete beginner to switch on a Mac and do something. This advantage disappeared when Windows became standard for IBMs and when the packages for the Mac became more and more complicated so that you now have to thumb through a manual two inches thick if you want to find out something quite trivial, like how to remove page numbers or to change the line spacing. But there was a period of time when the Mac meant status, presumably because of its high price, and this meant that nothing could stop them from sweeping through the biological laboratories of the world.

As an old-time computer user left over from the mainframe era, I found this very difficult. In fact, for years I had entirely the wrong idea about the purpose of the personal computer, for I still perceived it to be a calculating engine, and using a computer to be mainly about writing programs. I did actually write several programs—the most useful ones were probably those that enabled me to calculate the behaviour of my hypothetical gradients and to draw a number of the figures for the second edition of my book *From Egg to Embryo*. But nobody else ever wrote any programs, the reason being that computers are not any longer calculating engines at all—they are typewriters or terminals—and the commercial programs for doing all the things you do with typewriters or terminals are now so sophisticated that no amateur would any more dare write his own program than make his own radioisotopes. So when we were forced to change from PC to Mac, I lost all my programs and became a sour and embittered opponent of the Macintosh.

Once we had all become Mac users, we had to put up with the continual deterioration of our facilities, which is called *upgrading*. The process works something like this. The people in the central computing department need to keep their jobs, which means that they need to make themselves indispensable to the rest of us. Once they have installed and become familiar with something, whether it is an operating system, a communications program, a networking system, or even a word-processing package, they feel it is time to move on. Never mind that the users are barely aware of the existence of a facility, let alone how to use it properly. Suppose that a new operating system comes onto the market. Of course, we have to have it, and because all the institution's computers are networked together in complicated and overlapping ways, anybody who doesn't have it is going to run into trouble because he will soon have problems with compatibility with other users. So everyone has to have (and pay for) the new system. Then it turns out that various programs people are using no longer work with the new system. Sometimes they can be individually upgraded, sometimes not. Data collected in old formats may not be readable by the upgraded programs. It is all just a nightmare. The only thing you can be sure of is that after each upgrade, you will have fewer facilities than before.

The other process that degrades our lives is upgrading of programs by the software houses themselves. Each version of a word-processing program has more bells and whistles than the previous one. This would be a real advantage if you actually wanted to publish a magazine using a whole variety of layouts, formats, and fonts. But all we biological scientists need to be able to do is type a manuscript. Because of the extra facilities, however, you need a bigger computer to run it. About five years after the Mac era started, all the first-generation Macs from our labs had gone, not because they didn't work, but because they were unable to handle the new operating systems or packages that did the extra things we didn't need. Of course, we were in no position to resist the upgrading of packages in the way that a domestic user could, because the whole system was linked together, and if one computer behaved differently from the others it would cause even more trouble. When I moved to the University of Bath, I moved back into an IBM-compatible environment. So now the Macs we brought with us are underdogs again, and when they break down they will be thrown out and replaced by PCs.

Nowadays molecular biology cannot be done at all without the programs that analyse and compare gene sequences, or the more complex programs that model molecules and their interactions in three dimensions, so computing really is fundamentally important. But when it comes to the more mundane sort of computing used to write manuscripts and letters, the progress is less obvious. We don't have many secretaries anymore because everyone types their own stuff. So the work that was done by secretaries is now done by more highly paid scientists. Manuscripts now go through innumerable drafts and are corrected and recorrected until they are actually in the envelope. But to judge from the turgid and tortuous prose found in most scientific papers, the standard of scientific writing has not actually improved from its condition in the days of carbon paper.

Technology Transfer

The politicians have gone on for so long about the importance of commercialising every discovery that all universities and research institutes now have technology transfer organisations waiting to exploit the cornucopia of new results as soon as they appear, or even a bit before they appear. There are also outside companies that make a good living by promising to facilitate the development of innovations made by universities and institutes, although they normally expect fat up-front payments and seem less interested if asked to work on a commission basis. I have little doubt that developmental biology will eventually produce results of far-reaching importance to humanity, whether it be direct, like growing human organs in vitro for transplantation, or indirect, like identifying new genes involved in cancer. But so far the profits do not seem to have flowed so freely in my vicinity as to make me feel that our work is anywhere near the stage of being applied science.

In the SDS we were supposed to be discovering new treatments for serious diseases. The people who gave money to the organisation believed this, and so did the clinical staff, who were those working in hospital units with real patients. But the basic science staff knew that their work was light years away from application and always felt a bit guilty about this. It was not that the money was being misspent. Most of the areas of science being pursued were exactly those that told us something about the molecular mechanisms of serious diseases, and this even extends to topics such as growth factors or the mechanisms of animal development. But it is one thing to know the mechanisms and quite another to do anything about it. The management was always trying to promote the growth of something called the "clinical-laboratory interface" and would sometimes put some basic scientists into clinical units to try to help collaborations flourish along this interface. Alas, it was just those scientists who had the best understanding of the yawning gulf between basic molecular biology and seriously ill people, and they were the most cynical about the much vaunted clinical-laboratory interface.

Anyway, in such surroundings it was not surprising that high store should be set by discoveries with a potential for commercial development. The society set up a technology transfer company to handle such things, which was always supposed to be on the threshold of making large piles of money by licencing the society's discoveries to drug companies. It did make a bit of money, particularly out of antibodies that could be used for diagnostic purposes and some computer software written by the society's staff. But the profits from these things barely paid for the technology transfer company's running costs. Jasper, the director of the company, was himself an ex-pharmaceutical company employee and therefore knew that the things he was dealing with were rather small beer. Although they might be useful in their own context, they were just minor sidelines and did not correspond to the real purpose of the organisation. He was therefore always on the lookout for new gene products that he might be able to patent. Anything that got patented, or that was even considered for patenting, was the subject of numerous memoranda to the director, mentioning the name of the scientist in question, and therefore a useful source of credit within the society.

Unfortunately, Jaspar did not often come near me because I was tainted by the fact that I worked with frog embryos. Jaspar could not believe that anything that came from frogs could possibly be of any use in human medicine. In vain I argued that the neurotoxin from the poison arrow frogs of South America was a substance of high pharmacological potency, that the frog *Xenopus* had been very useful for pregnancy testing, that witches had used frogs for thousands of years . . . all to no avail. But one day I had the last laugh. It arose from a collaboration with my friend John Field. We had isolated a new type of fibroblast growth factor from the frog *Xenopus*. In the United Kingdom you cannot patent anything that has already been published in a scientific journal. So decisions about patenting need to be made before publication. Before publication I had informed Jasper about this new sequence, but of course he wasn't interested because it came from a frog. But then someone in John's lab had done a Southern blot on mouse DNA with our new probe. A genomic Southern blot is a method of visualising a gene complementary to a particular probe in a complex sample of DNA, so it can tell you whether one organism possesses a gene similar to a known gene in another organism.

I can report in parentheses that I actually saw the very first Southern blot in the whole of history. *Southern* is not a geographical term: the method was invented by one Ed Southern who was working in the rather northern University of Edinburgh and was on the floor above me at the time as I was doing my Ph.D. I was in his office one day and he showed me a piece of X-ray film with a smudge on it. "That's the *E. coli* ribosomal genes," he said. "Yes, very interesting," I said, "Now could you sign this form please?" I had totally missed the significance of the smudge, and I little imagined that in a few years the "Southern blot" would be as ubiquitous as the chromatography column or electrophoresis tank in molecular biology labs across the world, and that derivative techniques called "Northern," "Western," "North-Western," and "South-Western" would also be in daily use by millions of students and postdocs.

Anyway, someone in John's lab had done a Southern blot using our probe on a sample of mouse DNA and seen a smudge, or *perhaps* seen a smudge, as scientific results are not always as cut and dried as we should like to believe. John knew immediately what to do. Although he didn't work for the SDS, he did have a lot of grant money from another important charity for serious diseases, the Serious Disease Club (SDC). He immediately convened a meeting of the commercial director of the SDC, Jasper, myself, himself, and the technology transfer liaison director for the university to discuss the new breakthrough. We explained that a new FGF had been isolated from *Xenopus* and that the smudge meant that there was probably a similar gene in mouse. If there was a gene in mouse, there would certainly be one in humans, as the human is much closer to the mouse than to the frog. This sentiment is not in any way a judgement on our political leaders; it is simply based on the fact that mammals diverged from amphibians about 400 million years ago, but p   rimates diverged from rodents more like 100 million years ago, according to the fossil record. The differences in sequence of their genes reflects these divergence times. If there was a gene in human, then it was a new growth factor and it ought to be patented, because some drug company interested in FGFs would undoubtedly soon come along and buy up all the rights to all types of FGF for a gigantic sum.

Everyone was most impressed. But there was a problem. It is difficult to patent a gene that is simply the homologue of a known gene in another animal, and our paper describing the frog gene had been published 50 weeks previously. Everyone turned to Jaspar. "Why didn't you patent the frog sequence?" asked the commercial director of the SDC. Jaspar mumbled something about it only being useful to treat frog diseases. "How different are the frog and mammalian FGFs?" they asked me. "The sequences are slightly different, but biologically they're the same. *Xenopus* FGFs have the same specific activity when tested on mammalian cells as mammalian FGFs." "Why didn't you patent the frog FGF?" they asked Jaspar again. He shrank into a corner.

Of course, it didn't matter. The smudge was unreproducible. The mammalian homologue of the *Xenopus* FGF was probably already known, although the difference between them was greater than usual. The main U.S. biotechnology company that was working on FGFs dropped them a few months later, having decided that they were not, after all, any use for the acceleration of wound healing. Maybe FGFs weren't going to make anyone's fortune just yet, but at least we had got the full benefits of those memos passing through the director's office mentioning "novel technologies," "patents," and "commercial positions."

The Ground Plan of Evolution

The Fly

I remember reading a horror story in my youth entitled "The Fly," later made into a successful movie. It is the story of a man who invents a method for the instantaneous transfer of matter from one place to another. He became so confident in its operation that he started transferring himself across his laboratory from the transmitter to the receiver. Unfortunately, one day there happened to be a fly with him in the transmitter chamber. When they arrived at the receiver, they had become partially mixed up! The man had the head of the fly instead of his own head, and a fly's leg in place of one arm. The fly had received the human head and arm in place of its own organs. The story ploughs on to inevitable disaster. But it makes us wonder whether such a substitution could really happen. In other words, are there equivalent parts in the bodies of flies and people? Is the fly's head like the human head, and is the fly's leg like the human arm? Actually, flies have six legs, and the three pairs are not the same, so if this disaster had happened for real, *which* of the fly's legs would have been substituted for the human arm? Such questions may seem fanciful, but millions of dollars of grant money are being expended at this very moment in trying to answer them.

Drosophila and Its Mutations

The fly in question is not the common house fly but the rather smaller and less conspicuous fruit fly, *Drosophila*. It first came to prominence as an organism suitable for experimental genetics through the work of the great American father of genetics, Thomas Hunt Morgan, working at Columbia University. Morgan actually began his career with a study of the regeneration of parts that had been amputated in animals such as worms and newts; he wrote a book on the subject, published in 1900. Rather wisely, he realised that nothing much was going to happen in regeneration during the twentieth century, and so he changed fields to the study of inheritance, or genetics as we now call it. In effect, genetics had been founded by Gregor Mendel, a monk at the monastery in Brno, now in the Czech Republic. With experiments on peas, he had shown that hereditary variations, or traits, could be attributed to individual hereditary factors that remained unaltered over successive generations. The different appearance of individual plants arose because they carried different combinations of these factors, but they could not be seen and remained hypothetical. Although published in 1866, Mendel's work had little impact until the end of the nineteenth century because it appeared in the *Verhandlungen des Naturforschenden Vereins in Brünnen* (*Proceedings of the Natural History Society of Brno*). This was a somewhat obscure journal, and although it was available in a few centres in the West, it was little read. It is perhaps the most horrific example in history of the perils of publishing your work in a low-impact journal! But once Mendel's work was "rediscovered," the nature of inheritance became a central area of biological research. The term *gene* was introduced in 1909 by the Danish botanist Wilhelm Johannsen and has been used in just about every subsequent paper in the biological sciences. Using *Drosophila*, Morgan showed that the mysterious genes were carried on the chromosomes and that they behaved as though they were arranged in a line, with the probability of separation of two genes by recombination proportional to the distance between them. These discoveries formed the basis of genetic mapping, and in 1933 Morgan received the Nobel Prize for Physiology for his work.

Most insects have two pairs of wings, but flies are peculiar in having only one pair, which lies on the second thoracic segment. On the third thoracic segment, where the other pair of wings would normally be found, lies a pair of small balancing organs called *halteres*. As long ago as 1915,

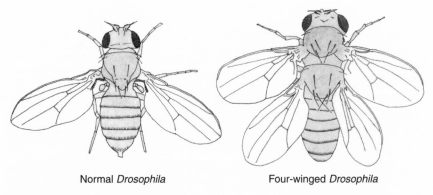

Normal *Drosophila*   Four-winged *Drosophila*

Figure 8.1. *Drosophila.*

one of Morgan's assistants, Calvin Bridges, had discovered a mutation that converted part of the haltere into wing tissue. He called it *bithorax,* because the flies looked superficially as though they had two thoraxes. Such mutations that convert one part of the body into another are known as *homeotic.* In Figure 8.1 is shown a drawing of a normal *Drosophila* adult, and also one carrying an *ultrabithorax* mutation, a more extreme form of *bithorax* that converts the entire haltere to a wing. This fly has four wings and is a truly remarkable beast. Until about 1980 hardly anyone had heard of homeotic mutations, and the subject was generally ignored. But a few developmental biologists had always realised that they must be of supreme importance. If a mutation caused a transformation from haltere to wing, then presumably the function of the normal gene in the normal animal was to convert wing to haltere. More precisely, it was to distinguish regions of the future second and third thoracic segments during embryonic development such that the former produced wings and the latter produced halteres. In other words, out of all the thousands of genes in the genome, homeotic mutations identified to us those few genes that had a direct role in the control of embryonic development.

Although it was not until the 1980s that it became possible to clone homeotic genes, many experiments on their biological effects could, technically, have been carried out at almost any time since 1915. They were not actually carried out until the 1970s and so immediately preceded the molecular work. The reasons for this are partly to do with fashion but also to do with a curious blindness afflicting all developmental geneticists of the period. One problem was that, if you have isolated a new type of mutant, there is a natural tendency to describe and think about the mutant rather than about the normal form and the role of the normal gene. But in fact, it is the normal function of the gene that is important; the mutant just shows what happens when the gene function is missing, reduced, or altered. So this caused a general slowness to ascribe normal functions to the genes identified by mutation. The other problem was that mutations were always studied in adult flies, and that meant that the flies had to survive throughout embryonic development, larval life, and pupation before they would even get looked at. But if the genes that control development are mutated, it is likely that development will fail at some stage and the flies will die as embryos, larvae, or pupae. Although for each gene there may exist some weak mutations that are capable of survival to adulthood, mutations in developmental genes are usually lethal, and that meant that most of them were never discovered.

These two misapprehensions, or misemphases of thought, delayed the opening up of an important area of science for at least 50 years. This is an interesting example illustrating that lack of progress in science is not simply a result of the lack of money, or the inadequate sensitivity of equipment—it may simply arise from conceptual blocks or limitations. Although the points may have been recognised at various times in the past, the first real response to it was the mammouth genetic screens for developmental genes carried out by Christiane Nüsslein-Volhard and Eric Wieschaus in the Max Planck Institute for Developmental Biology in Tübingen, Germany. This involved searching for mutations that not only killed the embryos but also produced some visible defect in the anatomy suggestive of a function in development. This comprehensive study led, in a few years, to the identification of most of the genes in *Drosophila* that controlled development, about 140 in all.

Homeosis

The term *homeosis* (then spelt "homoeosis") had been introduced by the British evolutionary biologist William Bateson in his book *Materials for the Study of Variation* published in 1894, 90 years before the work of Nüsslein-Volhard and Wieschaus. Like many nineteenth-century biologists, Bateson had difficulty accepting Darwin's theory of evolution by natural selection because there seemed no way in which it could operate. In those days, Mendel's work had already been carried out, but it was lying unread in the *Proceedings of the Natural History Society of Brno,* and most biologists in the West still believed that inheritance involved the blending of the hereditary factors in each generation. This is, of course, an observed fact. If children have one tall and one short parent, then their own final height is likely to lie somewhere in between. But it made it very difficult to understand how a novel advantageous mutation could spread through the population and eventually, as a result of natural selection, come to be the normal form of the gene. This is because, if genes are blended in each generation, then the advantageous effect of a mutation will be diluted out after only a few generations and there will soon be nothing left on which selection can operate.

We now know that genes retain their identity between generations but that characters such as height are determined by many genes, so the observed blending is due to small additive effects from many different genes. But in 1894 Bateson's response was to argue that the important mutations in evolution could not be the little ones with minor effects that would produce a gradual continuous change in form. They must instead be large ones with big effects that produced substantial change of form immediately, before the blending had reduced their effect too much. His book was an attempt to justify this belief by collecting an enormous zoological garden of naturally occurring abnormalities and to stress the discontinuous and abrupt character of the variants found. The collection included not only mutations but also a number of abnormalities that arise from regeneration of damaged parts. Some of these are also homeotic changes, such as the propensity of the crayfish *Palinurus,* after removal of an eye, to regenerate an antenna in its place. Others are not homeotic but are still very interesting, for example, the regeneration of multiple legs in insects following damage to the original leg. It is now possible to explain the number and arrangement of such supernumerary limbs by applying Lewis Wolpert's theory of positional information, but the molecular basis of their formation is still unexplained. For all these reasons, Bateson's book remains fascinating reading today.

By the early 1980s, the techniques of molecular biology were just becoming good enough to contemplate the cloning of genes that were known to have interesting mutations. The *bithorax* mutation, discovered early in the century, had not been entirely ignored, for it formed much of the life work of Ed Lewis, working at Caltech in Pasadena. He isolated a number of further mutations of the *bithorax* class and gradually realised that there was an interlocking group or "complex" of genes controlling the identity of the different body segments. Again, because he worked mainly with viable mutations rather than lethal ones, it was not until 1978 that he was able to publish an account of the effects of deleting the entire Bithorax complex. It was truly dramatic, as the absence of the gene complex caused all of the body segments posterior to the second thoracic to become transformed into further copies of the second thoracic segment. There are normally three thoracic and eight abdominal segments in *Drosophila,* so this mutant had no less than 10 copies of the second thoracic segment. Of course, the mutant could not survive long enough to become an actual fly; it was a dead embryo still inside its egg case. But the beauty of *Drosophila* for this type of work is that dead embryos are still useful to the developmental biologist. You can still see the general body pattern even in a dead embryo because the pattern of segments is revealed by the pattern of little bristles on the ventral side, which form at an early enough stage to be visible in most cases of embryonic death.

I remember reading this paper really carefully, as it was obviously a milestone in developmental biology. But it was an intricate paper full of genetic jargon, so even though it was published in *Nature,* few readers were mentally prepared for it at the time. My own reaction was to plug the results into my latest model, called the *serial threshold theory.* This model fitted a number of the results seen in the experimental study of regeneration and was also compatible with Wolpert's positional information theory, so I was pleased to see that it also fitted these new results of Ed Lewis. But there were a few people, more far-seeing than myself, and literate in both molecular and developmental biology, who realised that now was the time for the decisive application of the new molecular biology to the problems of development. The first task in this revolution was to be the cloning of the Bithorax complex and other important developmental genes in *Drosophila.* Ed Lewis was the third recipient of the 1995 Nobel Prize for Physiology, along with Nüsslein-Volhard and Wieschaus.

The Homeobox

The cloning of the Bithorax complex took several years, and by the end of this, it was clear that there were three genes in the complex and a large number of regulatory regions in the DNA in between the genes. The three genes appeared to be responsible for specifying three body regions, roughly the thorax, the front half of the abdomen, and the rear half of the abdomen. The next step came from the study of another rather similar gene complex called the Antennapedia complex. There had for some time been known a mutation called *Antennapedia,* which, as its name suggests, converted the antenna into a leg, and it had become clear in the early 1980s that this was part of another gene complex containing several genes that distinguished different parts of the head and thorax. In 1983 workers in two labs, Matt Scott at Indiana University and Ernst Hafen, Mike Levine, and Bill McGinnis in Basel, Switzerland, found a DNA sequence that seemed to be common to genes in both of the two complexes and to certain genes elsewhere in the genome. As the news filtered round the labs of the world, there was tingle of excitement and anticipation because this was one of the great discoveries of molecular genetics. The two labs had, independently, discovered the homeobox.

The homeobox is a sequence of DNA, 180 base pairs long, hence coding for a peptide of 60 amino acids, that makes up a part of many genes. It codes for a region of the protein that is specialised for binding to DNA. All homeobox-containing proteins are transcription factors: that is, they belong to the class of proteins whose job it is to control the activity of other genes. The homeobox received its name because it looked initially as though it might be a sequence characteristic of homeotic genes. In fact, not all homeobox-containing genes are homeotic, and not all homeotic genes contain homeoboxes, but the name has stuck because it has great power over the imagination. The real significance of the homeobox is that it is not confined to *Drosophila*. It was quickly looked for, and found, in the DNA of other animals, including vertebrates. This suddenly meant that the molecular biology of *Drosophila* assumed a massively greater importance than it had previously enjoyed. No longer was it just an issue of the mechanism of development of one small insect. Now it was suddenly realised that bits of DNA, identified as being of importance using *Drosophila* genetics, could be used as probes to identify and isolate developmentally important genes from vertebrates as well. This expectation was sustained, and after a period of confusion, it became clear that all multicellular organisms—animals, plants, and fungi—contained homeobox genes and that they are usually, but not always, concerned with development. There was one subset of homeobox genes in particular that came to attract an extreme degree of attention. These were the direct homologues of the Antennapedia and Bithorax complexes of *Drosophila* and are now known as the Hox genes.

Hox Genes

Hox genes are characteristic of animals rather than plants or fungi. In most animals, there is a single cluster of Hox genes ranging from 5 to 13 in number, although vertebrates with their more numerous genes show a more complex pattern, with four related clusters. *Drosophila* itself eventually turned out to be atypical, as its Hox cluster has become split into two separate groups of genes relatively recently in evolution, which are of course the Antennapedia and the Bithorax complexes. The importance of the Hox genes is that they control the pattern of body structures in the all-important anteroposterior, or head-to-tail, axis of the body. Every group of animals that has been examined, from sea anemones to humans, contains the Hox genes. In all cases where it has been possible to examine the parts of the embryo in which the genes are active, the same rule of "nested expression" is followed. This means that all the genes are on at the posterior

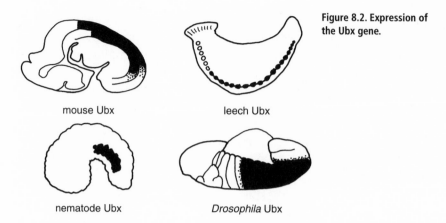

Figure 8.2. Expression of the Ubx gene.

mouse Ubx   leech Ubx

nematode Ubx   *Drosophila* Ubx

end of the body, and for each individual Hox gene, the domain of expression extends anteriorly to a particular characteristic body level. As Figure 8.2 shows, the expression of Ubx, one of the Hox genes, is very similar in animals as diverse as the nematode worm, the leech, and the mouse. In most cases the order of Hox gene expression from anterior to posterior in the body is the same, or nearly the same, as the physical order of the genes on the chromosome. This is a most remarkable fact, although the reasons for it remain even now somewhat obscure.

Each different individual gene in the Hox cluster has a different anterior boundary of expression. In three types of animal—*Drosophila,* the mouse, and the nematode *Caenorhabditis elegans*—it is possible to examine the effects of mutating Hox genes to inactivity. In all cases loss of a gene causes a homeotic transformation in which the most anterior part of its expression domain is converted to the structure normally lying anterior to it. For this to happen must mean that the Hox gene in question is the only thing distinguishing the two territories, so that when it is removed, the two territories have the same genetic identity and develop in the same way (Fig. 8.3). For example, in *Drosophila* loss of the *ultrabithorax* gene converts the first abdominal segment into a second copy of the last thoracic segment. Likewise, in vertebrates loss of the Hox gene having its anterior boundary of expression at the border between thoracic and cervical vertebrae leads to conversion of the first thoracic vertebra into an additional cervical vertebra. The converse effect is found if one of the genes is experimentally overexpressed. If a Hox gene is turned on in regions of the embryo where it is not normally on, the opposite type of homeotic transformation will occur such that regions of overexpression become posteriorized (see Fig. 8.3).

Control of anteroposterior patterning by the Hox gene system reveals an extraordinary conservation of developmental mechanism across the whole

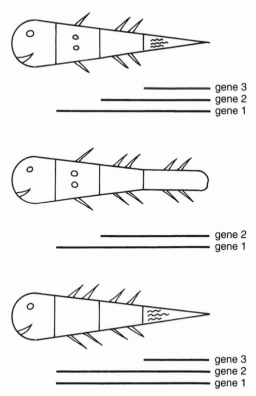

Figure 8.3. Homeotic mutations. The top animal is normal, the middle one has a mutation in gene 3, and the bottom one has ectopic expression of gene 2.

animal kingdom, and such conservation shows that all animals must, once upon a time, have had a common ancestor. This ancestor had Hox genes and used them to set up its anterior to posterior pattern of structures. My own interest in the Hox genes is twofold. First, as mentioned in Chapter 5, it has turned out that regulation of the posterior Hox genes is one of the main functions of the FGFs during vertebrate development, so the way in which this works has been a particular concern of mine. Second, the evolutionary and taxonomic implications of the Hox genes led me to write the "zootype" paper described in Chapter 4, which may not have changed the course of science but was fun while it lasted.

Homology and Analogy

It may not be obvious why the possession of a particular set of genes doing a particular job immediately leads us to conclude that all animals have a common ancestor. Might it not be that these genes are particularly well fitted to make the anteroposterior pattern and that all animals have thereby evolved a set that look more or less the same? This question has been asked many times in evolutionary theory, and classical zoologists have long drawn a distinction between two types of similarity between different organisms. They are called *homology* and *analogy*. If two structures are homologous, it means not only that they look similar but also that this similarity is due to descent from a common ancestor possessing the ancestral version of the part in question. The standard example for homology is the tetrapod limb. Many types of vertebrate, from humans to crocodiles, have limbs that are obviously similar in the number and positions of the bones and muscles, and in their general position on the body. There is every reason to suppose that they are similar because there was once an ancestral tetrapod from which all existing tetrapods are descended, and its limbs looked like this. Some biologists would also insist that homologous structures should be formed by similar developmental mechanisms, although there is not universal agreement on this as a criterion for homology. By contrast, if two structures are analogous, it means that there is no common ancestry and the parts look similar simply because the pressure of natural selection has forced a convergence of structure to meet the need for a similar function. A classic example is the insect wing versus the wing of a bird. Obviously, these do not have a winged common ancestor, because we know from the fossil record that there was a long history both of vertebrates and of insects before any winged members appear in either group.

When we consider genes rather than anatomical body parts, it is very easy to see whether they are homologous or not. This is because the protein products are composed of long chains of amino acids, and only a minority of those amino acids are usually necessary for the protein to do its job. The others may vary: in some positions there may just be a few of the 20 possible amino acids allowed by the constraints of achieving a functional protein; in other positions maybe any of the 20 will be tolerated. We now know that the sequences of genes and proteins change gradually over evolutionary time. Most of the changes are "neutral"—that is, they do not affect the survival or reproduction of the individual organism, and they are neither favoured nor discriminated against by natural selection. For any individual mutation, there is a very small chance that it will eventually spread through the whole species and become the normal version of the gene or protein. Because evolutionary time is very long, even a very small chance becomes in some instances a certainty, and there are many substitutions of amino acids in those positions in the protein where they do not compromise the biochemical function. The longer the time since the divergence of two lineages, the greater the number of differences in the sequences between them. But many inessential positions also remain the same, and it is the presence of these identities that show that two genes from different organisms are really homologous. Had the genes converged to the same function, they may have no identities at all, or any they did have would be confined to a small region of the molecule actually responsible for its catalytic or other biochemical activity. A molecular example of analogy is provided by the crystallins, which are the proteins making up the lens of the eye. In different types of vertebrate, quite different sorts of protein have been brought into service to make a transparent lens. Although they serve the same physiological function, they have no primary sequence in common. The different crystallins are analogous and not homologous.

To return to the Hox genes, we know that they are homologous because there are many similarities in their sequences that need not be there if all the protein has to do is bind a specific site on DNA. In fact, there are four clusters of Hox genes in vertebrates, whereas in invertebrates there is only one. The sequence comparisons strongly suggest that all four vertebrate clusters are of a similar evolutionary age, and it is likely that the four clusters arose from the single ancestral cluster at about the time of the origin of the vertebrates. Some brave souls would even suggest that the quadruplication of such important genes was the key event that *led* to the creation of the vertebrates.

The reason for believing that the cluster as a whole has always been responsible for specifying the anteroposterior axis of animals is that Hox genes are transcription factors. The function of transcription factors is to bind at specific sites to DNA and turn on various other genes. Transcription factors are to a large extent interchangeable. They are just like switches. One switch will work as well as another. It is not the mechanics of the switch that is important, it is its position in the regulatory circuit. Likewise with transcription factors. It does not really matter which one does a particular job; what is important is the wiring of the genetic regulatory network. For this reason it is inconceivable that if more than one lineage of animal evolved a head-to-tail axis, the same group of transcription factors would be chosen to specify the anteroposterior levels of the body. But the same group *is* used in all animals, so this means that all animals must be derived from a common ancestor. This ancestor had a Hox cluster and used it to specify its head-to-tail anatomy.

"I Spy" Is Not a New Game

In the life cycle of the fly, the egg hatches as a larva, which crawls around feeding and moults a couple of times before becoming a pupa. In the pupa the whole body is extensively remodelled. Most of the structures of the larva degenerate and disappear and are replaced by new adult structures grown from the imaginal discs. Imaginal discs are little buds of cells that are formed during early embryonic development. They grow to some extent during larval life and differentiate to form the adult structures during pupation. The imaginal discs, and some associated structures called abdominal histoblasts, form the entire external surface of the body in this way. Among the imaginal discs is one that forms a large part of the head and is called the eye-antennal disc.

The eye of *Drosophila,* like that of other insects, is a compound eye. This means it is formed from many individual eyelike structures joined together. Because there are many lenses and many little receptive fields, the compound eye does not generate a single image in the same way as the vertebrate eye, but nonetheless the insect brain is wired up in such a way that it can make productive use of the visual information it gathers. Because the compound eye is so different from the simple camera-like eye of vertebrates, these two structures have long been considered an example of evolutionary analogy. In other words, it has been presumed that they have arisen quite separately in evolution so that, however far back you go, there will be no creature with an eye that is a common ancestor to both vertebrates and insects.

Although everyone who works on *Drosophila* does so because of its good genetics, this does not stop them occasionally doing a bit of searching for genes using molecular biology methods to find homologues of known genes from vertebrates. There are in vertebrates a group of genes called *pax* genes. They are transcription factors containing a sequence called the *paired* domain. Following their success with the original discovery of the homeobox, in 1994 the laboratory of Walter Gehring at the Biozentrum in Basel, Switzerland, did a library screen on *Drosophila* using mammalian *pax* probes. One of the vertebrate pax genes, *pax-6,* had recently been identified as the gene mutated in the *small eye* mutant of the mouse. So Gehring's group was extremely interested when they found that the *Drosophila* homologue of *pax-6* was already known as a mutant called *eyeless.* In this mutant the eye-antennal disc in the larva failed to develop properly, and the adult fly had no eyes. So, contrary to all expectation, here were two genes that, from their primary sequences, had to be homologues, one of which was involved in making the eye in vertebrates and the other in insects. Even more spectacular, Walter Gehring's group then showed that if *eyeless* were overexpressed in other imaginal discs, the ones normally forming the legs and wings of the fly, these discs would instead form extra eyes. This experiment worked not only when the *Drosophila* gene *eyeless* was overexpressed in the other discs, but also when the mouse gene *pax-6* was overexpressed in the other imaginal discs of *Drosophila*! Not only were the genes homologous by the criterion of primary sequence, they had also conserved the same biochemical function over hundreds of millions of years of evolution in different lineages. Zoologists reconsidered their views about the relationship between the insect and vertebrate eyes and decided that there must have been an ancestral organism that had some form of photoreceptor, probably one without the advanced image-forming capability of the modern vertebrate and insect eyes, and that the *eyeless/pax-6* gene had been involved in its formation right from the beginning. So now we knew that the ancestral animal had a head and a tail and also had eyes.

The Heart and the Great Inversion

A similar story then unfolded in relation to the heart. In *Drosophila* there was a mutant that failed to form a heart. The gene was called *tinman* after the character in *The Wizard of Oz*, whose greatest wish was to have a heart. Cloning and sequencing of the gene showed that it was a homeobox-containing transcription factor—not one of the Hox genes, but one of the many non-Hox genes that also contain a homeobox. *Tinman* had a vertebrate homologue called *Nkx-2.5*, which was expressed in the heart during its early development. Mutation of *Nkx-2.5* (in mouse) could disrupt formation of the heart, and overexpression (in *Xenopus*) could increase the size of the heart. So, despite the totally different anatomy of the insect and vertebrate hearts, they seemed to have a least one key gene in common when it came to development. It rather looked as though the ancestral animal, unlike the original tinman of the movie, did have a heart, or at least some form of contractile element in its circulatory system, which needed the gene *tinman/Nkx-2* for its development.

It really was beginning to look as though all animals were the same, or at least that the developmental mechanisms for the formation of the main body structures of animals had been conserved over unimaginable lengths of evolutionary time. We began to doubt whether anything was actually analogous at all. Perhaps even the wings of birds and insects were homologous? But it still seemed that there was one big difference between vertebrates and insects: vertebrates have all the important parts on the dorsal side and insects have all the important parts on the ventral side. This fact had been discussed by the great French anatomist Etienne Geoffroy St. Hilaire back in the early nineteenth century. He had noted that in arthropods, which is the group containing the insects as well as other segmented invertebrates such as crustacea and spiders, the principal nerve cord and the main segmented muscles were on the ventral, or lower, side. The corresponding structures in vertebrates lie on the dorsal, or upper, side. Geoffroy believed that there was an archetype of animal form. He argued that even creatures as apparently different as vertebrates and arthropods could be considered to relate to this archetype, if it were admitted that one of the groups was turned upside down. The idea was soundly derided at the time by Geoffroy's arch opponent, the anatomist and systematist Georges Cuvier, and has been repeatedly dismissed by generations of evolutionary biologists from the time of Darwin.

But the derision has not extended far into the AC era. There is a gene in *Drosophila* called *decapentaplegic*. It was cloned in the 1980s and first attracted interest because the sequence showed that it clearly belonged to the TGFβ superfamily of growth factors. Mutants of *decapentaplegic* were defective in dorsal structures, and after much work, it became clear that the gene was active in a strip along the dorsal side of the embryo and that the protein product was secreted by these cells. Strips of cells in more ventral positions secreted an inhibitor coded by another gene called *short gastrulation,* and the combined action of the two genes led to the formation of a gradient of the active decapentaplegic protein from dorsal to ventral. The active decapentaplegic protein then induced the expression of other genes depending on its concentration, and this led to the main distinctions between the different parts within the dorsal half of the body. As seen in Chapter 5, it has been shown by several labs over the last few years that in the frog *Xenopus*, BMP-4—also a member of the TGFβ superfamily and the closest vertebrate homologue of *decapentaplegic*—is active on the ventral side, and its function is to promote ventral development. Various substances secreted by the Spemann organizer promote dorsal development by local inhibition of BMP-4. One of these, identified by the lab of Eddy de Robertis at UCLA, is known to be a molecule called chordin, and the gene for chordin in *Xenopus* turns out to be the homologue of the *short gastrulation* gene in *Drosophila*. Chordin works by binding to and inhibiting the action of BMP-4, just as the short gastrulation protein binds to and inhibits the decapentaplegic protein. So the system in insects and in *Xenopus* really seem very similar except that they are inverted with regard to each other. The BMP type factor in *Drosophila* is dorsal and its inhibitor is ventral, whereas in *Xenopus* it is ventral and its inhibitor is dorsal. These relationships surely argue for the correctness of the old idea of Geoffroy. Putting this into evolutionary language, we should now say that there was a common ancestor of vertebrates and arthropods and that at some time after the common ancestor stage, the precursor of one of the groups turned upside down and has stayed that way for the rest of evolution (Fig. 8.4).

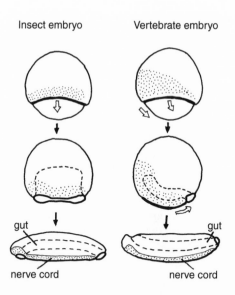

Insect embryo   Vertebrate embryo

Figure 8.4. The great inversion (after Arendt and Nübler-Jung, *Nature* *371* 26 (1994)).

gut

gut

nerve cord

nerve cord

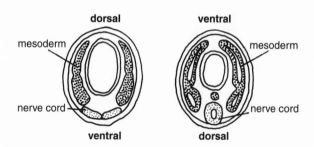

dorsal   ventral

mesoderm   mesoderm

nerve cord   nerve cord

ventral   dorsal

Lessons from the Fly

The story of the fly tells us many important things about scientific re-search in general. First and foremost, it shows that a problem that appears at first sight to be absurdly abstruse, like why a mutant fly has four wings instead of two, can, if successfully followed up, come to have a vastly wider significance. In fact, it has revolutionised our ideas about the struc-ture and development of all animals. Second, it shows that the study of the molecular biology of development can tell us a lot about what happened in evolution and about the nature of our long-dead ancestors. This in-cludes information about the anatomy of ancestors for which there is no fossil record remaining, something that we have always believed beyond the range of human knowledge. Third, although I have not described this in my narrative, it shows how quickly an area of science can be mined out.

Back in the 1970s, there were only a handful of people working on homeosis in *Drosophila,* and it was an exceptionally obscure backwater. Twenty years later generations of empire building have occurred. Great chiefs have risen up, acquiring mountains of grant money and numerous minions. Many of the minions have themselves become chiefs and formed empires of their own. Now, many of the second generation of minions are setting up new labs. Probably most universities in the United States have one or more labs working on *Drosophila* development, and they are now producing good time-expired postdocs by the hundred. When the cloning of the developmental genes started, a lab head could select one out of dozens of interesting-looking mutations to work on. Within a few years, it was hard to avoid competing with several other labs in the race to clone one of the few remaining plums. By now the early development of *Drosophila* is pretty well worked out in molecular detail, and the good labs have mainly shifted to the later, imaginal disc, stages. Some have diversified to other organisms, either to follow up the evolutionary implications or to apply some of the *Drosophila* magic to the more complex development of vertebrates. The lesson for an aspirant young scientist is that if you go into a field that is already booming, you are in for a tough time with the competition, whereas if you are lucky enough to start such a field yourself, you have at least a few years of fun before the competition catches up with you.

Paying for It All

There is never enough money for research. It does not matter if the amount spent now is 10 times, 100 times, or 1,000 times as much as what is was 50 years ago; it is still not enough. The reason bears no relation to the needs of society or the tasks facing the current generation of scientists. The reason is simply the impossibility of sustaining exponential growth for more than a few years at a time. Suppose that a good fairy decided to grant the wish of every academic in the world for adequate research funding. How much is adequate is a matter for individual preference. But for the sake of argument, let us consider the very moderate level of one graduate student and one postdoc. After three years the graduate student will get his Ph.D. and look for a postdoctoral position. The postdoc will look for an independent academic position. If he gets it, and acquires "adequate" support for his research, he will need a student and a postdoc to assist him. So even at this very modest level of wish granting, the level of support would double every three years as new labs were created to accommodate all the finishing postdocs.

Many academics consider an adequate level of support to be somewhat greater than this, say two technicians, three graduate students, and six postdocs. Postdoctoral positions normally last three years. So in this case, if all the postdocs got their own labs, the number of labs, and therefore the demands on the career and funding systems, would double every six months. The properties of exponential growth mean that it would not take that many years before the number of labs in life sciences was so large that it consumed the entire gross national product of the world. A few more exponential cycles, and all parts of the known universe would be incorporated into someone's lab.

Of course, a system with such growth potential can never be financed adequately, and this leads to the familiar perception of massive and worsening famine by the participants. In practice, as exponential growth cannot be sustained, many people have to leave the system to keep its size under control, and leave they do. In the United States the present situation is perhaps a bit worse than in Britain because there was a general expectation in the 1970s and 1980s that more Ph.D. graduates would be required to replace faculty who were due to retire in the 1990s. Graduate schools were very pleased to crank up their activity and generate the many thousands of Ph.D. graduates that the nation appeared to require. Alas, the end of mandatory retirement in the United States has meant that the oldies are hanging on to the bitter end and are not, after all, relinquishing their posts to the new generation. Accordingly, members of the new generation have had to make some difficult choices. Once they discover that they cannot get a research job in a first-rate institution, such as Harvard or UCSF, they decide to settle for a second- or third-rate one, perhaps a lesser-known university in the South or the Midwest. When all these become filled up, they turn their attention to the colleges. Liberal Arts colleges are institutions that have no real equivalent in Britain but are quite numerous in the United States. They are usually private colleges that teach undergraduates but do not have graduate schools and do limited research. They often have good reputations for teaching and are popular with students, mainly because they are smaller than the big universities and allow more individual attention from the staff. They are likely to have a curriculum in which the first two years are quite general and include topics such as "Western Civilization" (defined as DWEM, or the activities of dead white European males). They are dismissively referred to as "teaching colleges" by those in big universities, but recently, more and more postdocs have been beating a path to their doors in search of jobs. By now the colleges have filled up sufficiently that they are becoming picky about their new appointments, and a job is by no means available when you want it.

The other main escape route in the United States is the biotech company. There seems to be no shortage of these, partly because every academic in a good institution has his own company on the side, and partly because, as we saw in Chapter 3, in the United States, there is a limitless supply of private investors willing to invest in anything, however crazy it might seem. Every so often the biotech boom has a hiccup, as someone notices that the real income of the sector is rather small relative to what has been invested in it. But hopes of dramatic new cures, and of even more dramatic new profits, spring eternal; and it is likely that biotech will go on growing and that it will continue to accommodate the principal flow of biological scientists out of the U.S. graduate schools.

In the United Kingdom, Ph.D.s take only three years, rather than at least five in the United States. So Ph.D. graduates are perhaps more willing to settle for jobs outside academia. If they embark on a postdoc, they are making a risky choice, because nobody in Britain has any illusion that academic jobs will ever be numerous or easy to find. The escape routes for a time-expired postdoc are fewer than in the United States because there are few teaching-only colleges, and the biotech sector is still quite small. But whatever the country and the career opportunities, and however numerous the mouths that need feeding with jobs and grants, the majority of graduate student and postdocs still firmly expect that one day they too will become lab heads in their turn. Persuading people to adjust their expectations in a sideways direction is one of the responsibilities of the lab head, but it is hard work, and many people stay longer in the system than is really beneficial to their own prospects.

Careers in Academia

Around the turn of the century, the waiters in prestigious hotels and restaurants often received no salary. In fact, they might often have to pay or bribe someone to be allowed to work there. Their reward came from the tips that were expected to be so large as to add up to a better living than could be obtained by a paid post at a lesser institution, such as a cafe with only local clientele. The inheritors of this system are not the waiters of today but the biomedical scientists! Particularly in American private universities and in medical schools—and most of all in private medical schools —the faculty are not paid. When they are hired, they are allowed to occupy space in the department, but they have to find their salary, along with all their research costs, from external grants. Even when they get tenure, this does not come with any money but consists of being allowed to continue to occupy their space and to continue to write grant applications to support themselves and their research staff. Such is the lure of being in a good place that few younger scientists object to this system. After all, they have confidence in their own ability, so there should be no trouble raising a bit more on their grants to pay for their salaries. The system becomes somewhat more trying in midcareer, when people usually have families to support and their grant-worthiness, and incidentally their chances of moving elsewhere, is beginning to fade. It has also caused some political objections, mainly from representatives of U.S. states in which there are no private medical schools. Because most of the grants that feed the system come from the National Institutes of Health (NIH), which is funded by federal taxes, the people of states without private medical schools find themselves paying taxes to enable schools in other states to maintain their faculty at no cost. Meanwhile, they are also paying state taxes to maintain their own state medical school!

Things tend to be somewhat less harsh in state universities, as the faculty are usually paid 75 percent of their salary by the institution and are expected to find just the other 25 percent from their grants. This 25 percent is nominally the "summer salary" for the long summer vacation in which there are no teaching duties. But even in state universities, administrators have noticed the attractions of not paying their staff and, now that they are in a buyer's market, may make offers to newcomers at less than the 75 percent standard rate.

The normal career path in a U.S. university starts with appointment as an assistant professor. For those lucky enough to get a post, this would normally be at some age around the mid-thirties, after a Ph.D. and a couple of postdoctoral fellowships. Grants are hard to come by until the new principal investigator has proved himself, so universities often advance quite large amounts of money for setting up. At the time of writing, $250,000 is the going rate in good institutions. This money is used to buy equipment, hire research staff, and maybe a secretary, and to pay for consumables and the numerous charges imposed by the institution for animal care, washing up, and so on. If the post is one of those without a salary, the setup money might also have to provide for this until the grants start flowing. Once the grants do start, the university may want the setup money to be paid back, as well as taking its pound of flesh in the form of "indirect costs," about which more below. After five to seven years, the new faculty member will be assessed for tenure. If successful, this usually means promotion to associate professor and a permanent contract of employment. However, in especially prestigious institutions, associate professors may not have tenure and are, of course, most unlikely to receive salaries. Tenure in American universities is a major hurdle, and to jump it requires evidence of dynamic performance in grant getting and lots of publications in fashion journals. Teaching ability is also a consideration, although faculty at "good places" are unlikely to do very much teaching, about 30 contact hours per year being normal. Despite all the angst surrounding the tenure process, the majority of applicants do end up getting tenure, after which they are very unlikely to be thrown out. Promotion to full professor is likely to be achieved in due course, although it is by no means automatic. For those lucky enough to receive salaries, the 75 percent remuneration would be something of the order of $45,000 per year as an assistant professor, rising to maybe double for a full professor, although salaries vary quite a bit between institutions and individuals.

In the United Kingdom university staff are paid much less, but in some respects they have a better deal. Appointment is as a lecturer (salary about $30,000). Lecturers have a three-year probationary period during which they can be dismissed for poor performance, particularly in teaching. But if they pass this modest hurdle, their jobs are fairly safe. Failure to get research grants is less likely to result in dismissal than in being told to teach "Biochemistry I for Dentists," plus some significant administrative chores. Promotion to senior lecturer or reader is probable, although promotion to professor (salary maybe $60,000) is less likely. However, it is becoming easier to be a professor. There was a time not long ago when each department only had one professor, who was the chairman, or head of department. Now there are many "personal chairs," and in some departments it is expected that almost everyone will become a professor in the course of his career. Unbridled promotion does place a financial strain on the institution, and an imaginative solution to this problem has recently been found at Oxford, where people are now promoted to professor with no change in salary. This move has been necessary because the lecturers at Oxford naturally feel that they are so good that they would all be professors at a lesser university. In order that there should be no misunderstanding about their standing when they go to meetings, they need to have the title of "professor." Despite much fussing about standards and assessment, it is confidently expected that everyone at Oxford will be a professor within a few years. U.K. academics probably teach more than their American counterparts, about 80 contact hours per year being a normal sort of figure. But they have the advantage of actually receiving 100 percent of their salaries from the institution, rather than being told to find it from their grants.

Actually this is no longer quite true. In recent years the American system has spread to the United Kingdom in the form of the "research hotel." For many years there have been a number of senior fellowships offered by the funding bodies that enable independent investigators to do research full-time, without needing to be distracted by the teaching and administration that are the lot of a university staff member. The fellowships are for a fixed term of four to six years, after which the fellow must look around for another job (or another fellowship). Good university departments usually contain some of these externally funded senior fellows as well as their regular teaching staff. A research hotel is a place constructed specifically to attract such people, often a new building paid for by external funding, and is associated with some vague theme such as "molecular cell biology." Usually the only hotel occupant paid for by the university is the director, who will use his reputation and public relations skills to advertise the availability of space at this prestigious location. Thousands of energetic young people usually apply for these "spaces," and when their numbers have been whittled down to the actual space available, the lucky ones are told that they can now go off and make applications to the external funding bodies for their fellowships and other research support. The funding for such fellowships has increased considerably in recent years, mainly because of the growth of the Wellcome Trust, the biggest U.K. medical charity—although the applicants are by no means guaranteed success, and when their fellowships expire, they will have to leave and find a job elsewhere. Readers from other walks of life may wonder why on earth anyone would wish to work in a research hotel when the career prospects seem so bleak. But the advantage is that the fellow has no teaching and virtually no administration to do. He, or she, because senior fellows are often women, can devote 100 percent of their time to research in a very research-oriented environment. This means that they can build up a good publication record and perhaps get a better university job at some time in the future. If they play their cards right, they may be able to do a couple of senior fellowships, then move straight to a professorship in a university without bothering with the more junior grades of lecturer, senior lecturer, and reader.

Of course, a university has no chance of establishing a research hotel unless it is already a "good place." People want to be in good places and are willing to put up with minor inconveniences, such as no salary, in order to do so. For the university a research hotel is a fabulous advantage because it can claim all the publications of its members for inclusion in the Research Assessment Exercise, which is given every four years (about which see more below). The cost is negligible, as an external funding body will probably pay for the building and with a bit of luck, the core central services such as heating or washing up as well. The prospect of a research hotel can be guaranteed to make the vice chancellor (the executive head of a U.K. university) lick his lips in anticipation.

Sometimes incoming senior fellows try to bargain with the head of department along the lines of: "I'll come here for five years with my fellowship and thereby benefit your department with my funding and publications. In exchange I want a written guarantee of a lectureship at the end of it." Heads of department often try to give vague verbal reassurances and commit nothing to paper. In practice, the places that have big research hotels cannot make such commitments because they cannot generate enough positions for all the finishing fellows. Less good places that are just accommodating a few senior fellows may be able to do so and, in fact, may have to do so in order to entice the fellows to come.

Grants

In most walks of life, a grant is something rather demeaning. It is something you receive if you are not self-sufficient enough to earn money by providing useful goods and services to others. But in academia grants are the very lifeblood of the institution and are regarded as valiantly won to a much greater extent than individual's salaries (for those who get salaries). Scientific research is so expensive today that neither U.S. nor U.K. universities have the resources to finance research projects. So serious work in science cannot be undertaken without large external research grants to the faculty members. In the United States the source of grants in the life sciences is mainly the federally funded National Institutes of Health (NIH), together with the smaller National Science Foundation and some medical charities such as the American Cancer Society. In addition, the Howard Hughes Foundation has started supporting quite a few of the star players in recent years. In the United Kingdom grants come, in order of quantity, from the Wellcome Trust, government Research Councils, and various more focused medical charities such as the Cancer Research Campaign and the British Heart Foundation.

Grants come in various shapes and sizes. In the United Kingdom the most desirable type is the programme grant. This will run for five years and support several people. But more usual is the project grant, running for three years and supporting just one postdoctoral assistant. Both types are given by the major funding bodies. Grants are remarkably expensive. Using the cheaper U.K. figures to provide rough estimates, postdocs are paid about $25,000 per year, and the employment costs of social security and pension contributions add an extra 25 percent. Molecular biology consumables are very expensive; the annual consumable costs for each worker are about $15,000. In addition most grants will also include funds for some new equipment. So a three-year project grant is likely to cost about $150,000, and a five-year programme grant for four people over $1,000,000. But this is just the direct cost. Grants are also supposed to supply the hungry universities with indirect costs, otherwise known as overheads. These are supposed to pay for all the things the university provides that are necessary for a research environment, such as libraries, computing facilities, darkrooms, washing up, secretarial services, and so on.

Overheads may sound immensely boring, but they loom very large in university finance and internal politics. This is because the overhead money goes directly to the university rather than to the individual grant holder and can be spent on anything whatsoever. There is much acrimonious discussion about where overheads really go. Holders of grants often feel that the overheads are *their* money and should be spent on things that directly benefit them. Deans and department heads on the other hand like the flexibility of more core income and hotly defend their right to spend the overheads themselves.

In the United States the university calls the tune about the level of indirect costs, and the more prestigious private universities can command the most. As these are the institutions that do not pay their faculty, and who charge enormous fees to their students, one wonders what it is that they end up spending their money on! In Britain it is the funding body that calls the tune. At present the medical charities refuse to pay anything at all towards indirect costs, whereas the Research Councils will pay the sum of 45 percent of the gross salary bill. Because of the overhead, Research Council grants confer more status on the grant holder than other types of grant. However, grants from all sources are welcomed to some degree, as without them no research is possible, and without research the institution would sink into the Stygian darkness and become a "teaching university." Having suffered such a fate, it would be little consolation to reflect that that was what the public thought it was all along.

As all faculty in all universities are under intense pressure to obtain research grants, the pressure on the funding bodies is also very high. It is generally regarded as quite good if 20 percent of the applications are funded, although even those proposals that are funded will probably be cut back fairly savagely from the level of support requested. That large numbers of perfectly worthwhile and feasible applications are rejected at every session is the main cause of the widespread feeling that there is a desperate shortage of funding. To have any chance of success, grant applications must be very carefully prepared. Grants are awarded by panels, called study sections in the United States, that assess and compare applications at their meetings. All the applications are sent out to reviewers, who are academic scientists like the applicants, and their reports are available to the panel when they make their decision. Applications are assessed by how interesting they are, how important they would be if the work is successful, what chance the work has of being successful, the track record of the principal investigator, and how "good" the place is where the work will be done.

The main problem with the system is that nothing is funded unless it is guaranteed to succeed. Unlike research in the arts or social sciences, where some sort of result is bound to be forthcoming, even if not very interesting, experiments in the natural sciences may easily fail altogether and yield nothing at all. Of course, it is possible to design a proposal that is pretty well guaranteed to succeed, but then it will not be at all interesting. In this regard research is rather like investment on the financial markets: small rewards can be had for little risk, but big rewards mean very big risks indeed. Experienced applicants try to mix in some interesting high-risk stuff with some routine boring stuff. But most applications follow a standard pattern for the field in question. The reviewers know that what is proposed is like much of the current mainstream work in the field, including their own, and cannot bring themselves to give it a really low grade. On the other hand, they don't like to make it too easy for their competitors to get funded. So most reviewers' marks lie on the borderline between funding and nonfunding

Nowhere on earth has grade inflation bitten deeper than for the scoring of grant applications. The grading system differs for different bodies, but everyone knows where the cutoff comes. For example, U.S. grants never have a chance unless they are rated "outstanding," rather than "excellent," "very good," "good," "above average," or "average." In the United Kingdom, where we have a strange reluctance to use the extremes of any marking system, grants need to make at least alpha-4, where alpha-5 is the top, alpha-1 the bottom, and the remaining letters of the Greek alphabet represent the bottomless pit of degradation never entered by mortal scientists. The irony of this is that since only the work that is funded will get done, and since the quality of this will be much the same as what was done in the previous three years, the real grade of the funded applications is actually somewhere around "average." Most of the unfunded ones are probably also around "average," but were a bit less lucky. Because the reviewers' reports serve only to eliminate the really poor applications, the main job of choosing the lucky minority to be funded falls to the grant panel. In the United Kingdom at least some of the panel are likely to know most of the applicants personally, so the informal assessment of standing, competence, and innovative ability of the applicant is very important indeed. For this reason it does not pay to fall out with people sitting on grant panels or people who are likely to do so in the future.

The five-year programme grants, because of their size, are more susceptible to small-p political influence than three-year project grants. This mainly means an assessment of whether the applicant's field is a desirable one for the funding body to support. Is it an up-and-coming field or an unfashionable declining one? How would this grant contribute to the overall portfolio? To most people the portfolio of individual funding bodies must be one of the least interesting things in the universe. This is because it is almost invisible to outsiders, as many different bodies fund basic cell and molecular biology and most successful university scientists have grants from more than one source. The scientists, of course, like to have several sources of funding so as to spread the risk of failure more widely. Funding bodies do, however, take their own portfolios very seriously, and this can work either for or against the applicant. For example, if my good friend Professor Able and I are the only U.K. experts on, let us say, growth factors in fish embryos, and he has a big grant from the Medical Research Council (MRC), then I should be well advised to look elsewhere. If I send an application to the MRC, they will say: "How does this guy differ from Able, who we already have?" On the other hand the Biotechnology Research Council might say: "We aren't doing anything on growth factors in fish embryos. Slack isn't quite as glamorous as Able, but he is okay. It would fill a gap in our portfolio." But the grant-giving bodies also like to reward loyalty, a bit like supermarkets with their cash-back cards. So if you already have several grants in your department from one source, particularly if they are for similar things, then you can argue that this is a "centre of excellence" and "an important concentration of resources that can meet competition at the highest international level," and this helps to get another one. What the funding bodies like most of all is being "proactive," and supporting large projects that they can be clearly identified with. They would normally prefer to set up a special unit to, say, catalogue all the genes in the pig, or some other economically important organism, than to spend the same amount on a variety of different project grants in universities. This is a source of some irritation to universities in a situation of cash starvation.

The problem with being proactive is that it involves the research strategy and objectives being set by a committee of the great and the good. This has about as much chance of showing creativity and originality as a large canvas painted by a committee of art school directors. All too often an initiative that looks reasonably fashionable at the outset becomes in the end a long-lived and expensive institute full of tenured personnel doing rather little. Of course, in the more applied type of research, where the basic scientific understanding already exists and the desirability of a specific technological objective is clear to all, there is a lot to be said for setting up a specific organization to do it. But in the life sciences, private industry has always excelled at developing the actual marketable products, and there comes a point where it is more appropriate to transfer an applied programme to industry rather than attempting to complete it in the universities and institutes whose function is really that of extending basic knowledge and of training personnel.

The University

Members of the public, as well as some politicians, seem to think that the job of a university is to teach people things. But this view is rarely held by university staff themselves, who have undergone years of socialisation to condition themselves into believing that teaching is an irritating distraction from their real function, which is research. There are many linguistic giveaways to this. Teaching is frequently referred to as a "teaching load," and other (inferior) institutions are often referred to as "teaching universities." My institution is, of course, always a "research university," and so is yours, although perhaps not quite such a good one. In fact, this has almost become one of those "irregular verbs" beloved of the satirical BBC-TV (PBS) programme *Yes Minister:*

I am at a Research University

You are at a Good University

He, she, or *it* is at a Teaching University

In America league tables of institutions are constructed simply by asking deans and other senior figures their opinion and then averaging the results. This is subjective but cheap. In Britain there is a more elaborate system called the Research Assessment Exercise (RAE), which is carried out every four or five years by the government's Higher Education Funding Councils, and this has contributed further to the high status of research and the low status of teaching. The logic of the RAE is very simple. All U.K. universities, at least all the ones that you have heard of, are state universities, so they depend to a large extent on the government for their funding. Money for teaching is doled out *per capita,* and the number of students that any individual university is allowed to take is rigidly controlled by the government. So it is not possible to increase income by recruiting more students. But money for research is doled out very unequally, with those that have the most receiving the most. Although most research funding comes from external grants, the government money given direct to the universities is very useful money indeed because it is not tied to particular projects and, like the fabled overheads, can be spent on absolutely anything. So the prospect of a forthcoming RAE can drive vice chancellors, deans, and department heads into a sort of feeding frenzy as they vie to hire the best academic stars before the next deadline. The format of the RAE does tend to favour smaller and well-organized universities, and so by the time of the 1996 RAE, my new institution, the University of Bath, had managed to bootstrap itself up to a position just below Oxford, Cambridge, and London Universities, thereby augmenting its status substantially, at least within the United Kingdom.

Why the inequality in research funding? Although every academic in every post-high school institution in the world would like to do research, and would like to be given lots of money to do so, in reality there will never be enough money to make this possible, especially considering that doing research doesn't just mean doing experiments yourself. It means setting up a large lab employing a dozen people, most of whom will want their own independent position (and several assistants) in a few years time. So research support has to be rationed. Another reason for rationing is that research is international, and if something important has happened in California or Tokyo, British scientists will know about it within hours by phone, fax, or e-mail. So there is absolutely no point doing any research at all unless it is of international quality, and because the best lab heads in the United States and Japan are in good places with top-notch facilities, then ours must have the same; otherwise they can't compete. These two arguments force a high degree of selectivity in research funding.

So the purpose of the RAE is to find out, for every subject area at every institution, how good it is and to reward it accordingly. At least in the sciences, a lot of time could be saved by simply recording how much external research income was received and then paying a percentage of that to support indirect research costs. But this would not work for Medieval History or Russian, so everyone has to prepare a lengthy submission including the four best publications of each research-active staff member, plus information about external funding, and various free-format sections for saying how wonderful your department has been and how it is going to become even better. The assessment is made by a committee of the great and the good, who award each subject category at each institution a grade from 1 to 5*. In terms of research funding, a grade of 1-2 means ruin and 4-5* means being able to stagger on until next time. Even more significant are promotions and demotions. An improvement of grade means more money to play with, whereas a demotion means that someone has to be sacked.

About one to two years before the next RAE is to occur, tongues start wagging in the senior common rooms about removals and transfer fees. The key element in the system is that if an individual moves institution, he takes with him his publication list and grant funding, and frequently his postdocs, Ph.D. students, and equipment as well. Recruitment of one star professor can therefore make a significant difference to the appearance of a middle-ranking department, and the loss of the best research star in a department can well be serious enough to cause relegation to the next grade down. In reality, there is a lot more talk about transfers than there are actual transfers, because it is very tiresome to have to move across the country when it means disrupting children's education and giving up the spouse's job and circle of friends. However, it does happen and those who move know full well that it is their only real chance of ever getting any money out of an institution. During those critical few weeks when they have been offered the job but not yet signed on the dotted line, they have very considerable bargaining power.

This power is exerted to secure a fairly standard set of things that have high value in the academic world. One is space. Department heads have authority over space, and good places are always desperately short of space. Poor places on the other hand often have unoccupied space or can at least generate a large amount without inflicting too severe pain on those already in the department. This gives them an advantage. No incoming professor would accept less than 100 square metres, and might want up to five times this. Then there is refurbishment. Incomers like to have the lab laid out and furnished they way they want it and not how retiring Dr. Grey has left it. This can be expensive, so wise department heads will delay any refurbishments until they make a senior appointment. After all, the people already in the department are unlikely to leave just because they have to wait another few years for a cupboard or a new coat of paint. Next comes "setting up." This is a dowry of money paid to the incomer for equipment purchase. It is very useful money because it is untied in any way. Grant funds, which the incomer is likely to have a good supply of, are always highly itemised, and funding bodies can be awkward with people who change their minds about equipment purchases and be even less likely to cough up more if some unexpected problem develops. Although the sky is the limit in U.S. universities, lower expectations and the presence of more shared equipment means that $75,000 might be a usual starting offer to an incoming professor at a U.K. university. Naturally, it is fun to see how much higher you can drive it. Mind you, $75,000 is not very much with which to equip a modern molecular biology lab. This cannot be done for under $300,000, and one could quite easily spend $1,000,000 for a state-of-the-art facility. And there are specialised requirements like animal accommodation or greenhouses that may have to be added on to the laboratory refurbishment.

Then there is the knotty matter of *guarantees*. The incomer may need to make new grant applications to support some of his minions. Because these cannot be guaranteed to succeed, he will probably ask the new host institution to underwrite the salaries for a couple of years so that the people concerned can move knowing that they have a job to go to. Guarantees are a difficult area because they can be extremely expensive if called in. Department heads usually try to give verbal reassurances that can be reneged on if necessary, while the incomer will try to tie it all up in writing in a way that would stand up as a contract in court. Then there are the *positions*. These are junior faculty jobs in the gift of the incomer. The object of this is to enable him to appoint someone with similar interests so as to encourage the formation of a "critical mass" in the subject area. It is normal to offer an incoming professor one lectureship, so two or more represents a good deal, (although having just written this, I have just heard of someone who has got eight, so I am thinking of revising my mental scale of acceptability). These positions cannot usually be afforded as *de novo* creations, so they must be hoarded from the last retirements.

Finally, and of most interest to the wagging tongues, is the individual's salary. U.K. professors do not have a salary scale but are paid within a broad range from about $45,000 to $75,000. Anyone who is promoted internally will just go onto the bottom end of the range, but incomers can bargain for more. Very famous candidates can bargain for even more, and it is rumoured that figures as high as $120,000 have been paid in recent years, matching the sort of figure expected by a high flier in the United States. In this regard it is generally thought that universities other than Oxford and Cambridge have an advantage because it is often in the power of a single individual, such as the dean of science, or the vice chancellor, to make a decision about a salary and tie up the deal. In Oxford and Cambridge, on the other hand, such matters have to go through innumerable committees, and by the time a decision has been made, the candidate has decided to go elsewhere. Despite the advantages that the provincial universities may theoretically have in relation to space and salary, academics are so conscious of the pecking order of research status that no one would seriously consider going to a lower grade of institution almost regardless of what was on offer. After all, if you are at a good place, then people will think you are good. This means that you will attract good postdocs, they will do good work, and then you really will be good. The converse follows just as inexorably, and the conclusion drawn is that a drop in institutional ranking is a step on the road to ruin.

Once you have hired your superstar, the main problem is that he (it is usually, although not always, he) is always leaving. At any given time, there is an invitation in his briefcase to consider moving his magnificence to another institution. The offer involves an even bigger salary than he is getting already, vast setting-up funds, infinite space, maybe some core running costs, and no duties whatever except to *be there* and to benefit the host institution by the emission of beneficent rays of intellect and quality. So handling superstars can a bit of a headache for the department head, as they will always be wanting extra space and facilities, and of course, everyone knows about the job offers they are continually getting. It is important to realise that the superstar never actually *applies* for a job, going through the tiresome and demeaning process of competing with other applicants. By definition, superstars already have good jobs and do not need to look for another. That is why, if you want to hire one, you must *approach him* and make him an offer he cannot refuse. This at least is the theory. The practice may be a little different and involve some soliciting by the superstar to elicit these useful offers. But such activities will certainly only occur in the discreet corners of international meetings, well away from the glare of publicity.

An interesting example of RAE recruitment occurred when my friend John Field became fed up with the lack of opportunity for advancement in the Byzantine maze of Oxford and negotiated himself a professorship in the large provincial university of Central City. By then he was a very desirable recruit with several publications in *Cell* (worth lots of impact points in the RAE) and a large number of minions supported by two programme grants. The terms were very good, with massive space, huge dowry, appointments, an agreeable salary, and no teaching for four years. But for some time after he officially "moved," that is, started to be paid by Central City, he and his lab remained in Oxford as though nothing had happened. How admirable, I thought; now it is no longer necessary for universities actually to find space for their new recruits and it is no longer necessary for academics to move house with all the tiresome disruption to spouse, family, and friendships. Everyone can stay where they are, and universities can just bid to pay the salaries of the best stars. Modern computer technology will make it an easy matter to decide whose publications are attributable to which institution and award the RAE scores accordingly. Universities could dispense with all the complex apparatus of lecture halls, workshops, animal houses, and so on, and shrink to just a terminal in the vice chancellor's bedroom. He will play the academic transfer market in the same way that brokers can trade in millions of dollars worth of stocks or commodities without ever seeing them. The profits from dealing in superstars could then be used to buy distance learning services from the Third World, which would meet the university's teaching obligations at very low cost.

Alas, this attractive vision was not to be. Central City eventually somehow persuaded John to move, and after about a year, he was actually there in the flesh.

Portree

My own experiences of being courted are quite modest but still interesting in their small way. Once upon a time, I heard that some scientists of my acquaintance had announced plans to move to a remote northern univer-

sity. How nice, I thought, to work in such a beautiful place near the mountains! After making some discreet enquiries, I found myself invited up to
make a visit. I later found out that this place, the University of Portree, was
famous for its hospitality and that recruitment was taken very seriously indeed. At that time any half-respectable visitor was given a magnificent dinner, accommodated in a grand Scottish baronial hotel, and offered whatever
professorships happened to be vacant at the time. Biological sciences had
become the speciality of this place as they had realised quite early in the period of the Thatcher terror that survival depended on concentrating single-
mindedly on activities that brought in external funding. There are now considerable resources for the life sciences, mainly from the Wellcome Trust
and other medical charities, so life sciences had become a top priority.
Also, the University of Portree had already achieved a signal success. Its biochemistry department had been built up into a centre of international excellence by some very hard-working and enthusiastic individuals. It had an
enormous external income and dominated the rest of the university.

At dinner I was told how wonderful Portree was in all possible respects.
The scenery was incomparable, the climate ideal, the sports facilities could
not be bettered elsewhere, you could park your car anywhere you fancied,
and the schools were in a different educational universe from their be-
nighted cousins in England. I was told I could have one of two vacant
chairs; one was primarily a research post in the biology department and
the other was the headship of the anatomy department. I was advised that
the anatomy department wasn't very good, but it would be quite easy to
clear out all the existing staff and replace them with bright young persons
capable of pulling in enormous grants. Even at that stage, I suspected that
my hosts, who were from the biochemistry department, were not fully
conversant with anatomy teaching. The main problem with anatomy as
compared with biology is that the syllabus is dictated by the needs of the
medical curriculum and cannot be varied to suit the research interests of
the staff. Also, because most people who study anatomy are medical students, they can later earn three times as much by practicing medicine as by
teaching anatomy. This means that it is always hard to recruit good anatomy
teachers and very difficult to recruit someone who is a good anatomy
teacher and likely to be a success at research.

Being unused to flattery and its heady effects on the system, I retired to
bed in a state of high excitement, wondering what would happen the following day when I should meet the current anatomy staff themselves.
When the incumbent (acting) head of department turned up the next
morning, I realised I was in the company of quite a different type of individual. Professor Greybeard was a genial soul, and he explained to me why
the university was looking for a new head of department. First, the de-

partment had lost a lot of space to the predatory biochemists and had been almost entirely evicted from the new and palatial "Medical Sciences Institute." Their toehold indeed now consisted only of the dissection rooms, and they had managed to keep this only by playing their ultimate trump card of having to abide by the various legal requirements for the storage, handling, and transport of dead human bodies. In compensation for the rest of their losses, they had been given a crumbling old building in the centre of the campus. A vigorous new head of department was needed to supervise the conversion of this building into something useful. It was also necessary to recruit some more staff, as the teaching was only just being covered, and if someone left or fell ill, there would be a serious problem. Professor Greybeard confirmed my suspicions about the difficulty of recruiting anatomy teachers and said that in their last advertisement they had specified that "no previous knowledge of anatomy is required." For a teacher in higher education, such a stipulation was quite a giveaway. During the day I met all the other members of the department, who were very friendly. I gradually realised that the whole university lived under the shadow of the domineering biochemists, and that the anatomists were desperate to recruit a strong head of department who could take on the biochemists in the university committees and secure for them a reasonable share of money and space. At lunch I inadvertently referred to the Medical Sciences Institute as the "biochemistry department," and was left in no doubt that this was an unforgivable faux pas.

I was astonished throughout the visit that I should be, apparently, seriously considered for this post when my own knowledge of gross human anatomy was virtually zero. True, my research work was sort of anatomical because it dealt with how the major body structures became established in the embryo, but there is a yawning gulf between this sort of thing and being able to teach students all the names of structures running alongside particular nerves or blood vessels. I suspected that a department head who knew as little about the subject as I did would be absolutely at the mercy of his subordinates. If they wanted $100,000 for some piece of equipment absolutely essential for teaching the structure of, say, the chin, how would I be in any position to resist? It was also clear that the anatomy department had reached rock bottom. It would need total rebuilding to accommodate modern labs. It had no modern equipment, no facilities for working with radioactivity, and its animal house needed modernisation to comply with minimum legal standards. Worst of all, it had no administration. The work normally done by a departmental administrator was being done by Professor Greybeard.

Because the biochemistry department was a miracle, Portree had come to believe in miracles for the life sciences generally. The university was willing to put a bit of money into refurbishing the anatomy department, but they were really hoping that a dynamic new head would attract millions from a medical charity and that this would pay most of the costs. So the duties of the new man appeared to consist of the following:

Raising several million pounds for refurbishment

Establishing and running a dynamic research group capable of raising substantial grant finance

Administering the rebuilding of the department

Administering the day-to-day affairs of the department

Recruiting dynamic young lecturers who would also raise substantial grant finance

Taking his turn in teaching, examining, admissions, and all the other routine chores of university life

In short, they wanted someone to come and do about five full-time jobs for one salary. I began to understand the reason for all the flattery. Someone capable of all that would surely deserve it.

There was also the research professorship on offer in the biology department, conveniently situated opposite the biochemistry department (oops! I mean the Medical Sciences Institute). When I arrived for my appointment, I was greeted acerbically by the head of department, Professor Sharp. Before I could open my mouth, he said that he was not interested in the tentacles of the biochemistry department being extended into his domain, clearly regarding his vacant position less as a chair and more as a kind of Trojan horse. He then asked what on earth someone like myself, who already had a professorial level research job in *Oxford,* could possibly want in a poky hole-in-the-corner sort of place like *Portree.* Interesting. Had he been a biochemist, the idea that Portree was anywhere other than the centre of the universe would not have crossed his mind. I struggled to give convincing answers on these two difficult matters, and then the rest of the faculty trooped in to interview me. "What did I have to offer," they asked me. By then I knew the correct answer: "I can raise substantial grant finance," I said, and the atmosphere began to improve.

On reflection I decided that Professor Sharp was right and that there was no point exchanging my present job for a rather similar one in his department. I did hesitate over the anatomy chair for a while, being attracted by the flattery and the illusion of power it offered. But I decided that these things would wear off very quickly and then I would be faced with a hard grind of a job that could not possibly be done by one person, and that might well cut me off from research for good. So in the end, I waited. In fact, I waited until the SDS ran out of money and then I had to move. By this time Portree had demonstrated the effectiveness of its methods to the whole world. It had bootstrapped itself up to a commanding position in the life sciences and even had its own research hotel capable of attracting good people from across Europe and the United States. People in good places were beginning to mention Portree in the same breath as Oxford and Cambridge. But by then it had become such a good place that it didn't need Slack any more.

The Case of the Headless Frog

Although scientists love the publicity arising from big papers in the fashion journals or from large and important plenary lectures at meetings, they can be very suspicious of, and indeed hostile to, the general media. It is felt that newspapers and television can't be bothered to get the facts right and are always interested in sensation rather than serious science. The worst sin in science, apart from outright fraud, is to release data to the general media before it has been published in a scientific journal. This is because the refereeing process operated by the journal will, as far as possible, guarantee the validity of the results. Refereeing is very subjective when it comes to assessing the "importance" of a paper, but it is quite effective at guaranteeing technical standards. By contrast, in the rough and tumble of a press conference, there is very little chance that the results will receive any serious and expert scrutiny. This may be why it is not uncommon for biotech companies to release news about significant milestones by press conference; the main audience is not other scientists, or even the general public, but rather the potential investors.

In Britain the social standing of a newspaper can be instantly judged by the size of the pages. Large pages means high-class, small pages, or "tabloid" format, means low-class. The tabloids used to be known as "the gutter press" and do have a general reputation for irresponsibility and sensationalism. Although scientists' cynicism about the media is probably justified in the case of the tabloids, the surprising fact is that journalists working for serious newspapers or for television companies do try hard to get it right. But there are huge differences between their world and the world of science. Even a large national newspaper will probably only have one science journalist, so he or she will have to cover everything from astronomy to zoology. He will know some of the language but cannot possibly appreciate the minute details that are the everyday currency of the scientists' lives. He will be working to a deadline so will need to establish the facts and write his story within a couple of hours, whereas the scientist may have spent years meticulously checking and rechecking his results. The journalist is writing for a general audience who will not be interested in the details of the work but will want to know all about its long term significance and likely practical benefits. Moreover, however careful the journalist is, the story he writes may be very different from what finally appears in the paper. The subeditors may cut it to fit into a particular space, and they, rather than the journalist, will write the headline. Similar considerations apply to television, and it is mainly because of the very different agendas of the scientific and media worlds that bad feeling can often arise when they are brought together.

The Call

I had always wanted to be on *Horizon. Horizon* is the only serious British television programme about science. Many of my contemporaries have been on it, and I wanted to be on it, too. My first chance appeared to come when they phoned me up about the "zootype." Journalists are very generous with other people's time, and on this occasion I talked to them for hours on the phone. I said all the things I could think of to make them come along and film, but the trouble was that the programme was really about evolution and not about development, and I work on development and not on evolution. So they never came and filmed, and I didn't get to be on television.

A couple of years later, they called again. This time it was about Dolly. Dolly was the sheep cloned from an adult tissue culture cell at the Roslin Institute near Edinburgh; it had attracted a huge amount of media publicity, mainly because of the implication that if you could do it with sheep, you could do it with humans. There had been enormous speculation about possibilities such as parents wanting to clone a dead child, or husbands wanting to clone a youthful version of their aging wives. *Horizon* was making a programme about cloning and its implications. They had phoned me because, when the story first broke, the *New Scientist* had phoned me for a reaction and mentioned my name in the resulting article. The *New Scientist* had phoned me then because of the publicity I had had at the time of the zootype.

I tried to sell them the line that I actually believe, that the importance of Dolly was not simply the cloning of an animal but that it had been done starting from tissue culture cells. The procedure had been to enucleate a mature sheep oocyte and then fuse it with a single cell that had been grown in culture. The cells had originated from another adult sheep, and the cell line had been frozen in the interim. The ability to use cells was important because it would open the road to real genetic engineering of animals or, conceivably, of humans. This is because when you use cells, it is possible to select for rare events. If a gene is introduced into a population of cells, it will only go in properly in a few of them. If a gene is excised from a population of cells, this will only happen properly in an even smaller proportion. To recover cells with the desired genetic change, you need to be able to select just the ones you want from a large population, something possible to do with tissue culture cells in a way that can never be done with eggs. I explained that we now understood a great deal about development. For example, by altering one or two genes, we could suppress or duplicate parts of the body. So the ability to clone animals from tissue culture cells would open enormous new possibilities. In the end I uttered the fateful words: "Maybe if our knowledge of how to suppress body parts were combined with the cloning technology, we could grow human organs for transplantation." That did it. They were hooked. They would come and film.

I did feel a bit uneasy about it because my lab had nothing at all to do with the sheep cloning work or with organ transplantation. On the other hand, I felt they really ought to have a developmental biologist on the programme, and if I didn't do it then they would just go to someone else, who might know even less than me and do a worse job!

The Science

Our flagship project at this time was work on how the head-to-tail pattern of the embryo becomes determined. As mentioned in Chapter 5, we had shown that FGFs were expressed at the posterior end of the gastrula and that they turned on the Hox genes required to make the trunk and tail parts of the body. At the time of the *Horizon* phone call, this line of work, carried out by my postdocs Harv Isaacs and Betsy Pownall, was nearing completion. The story was that the FGFs turned on transcription factors of the Cdx class, which in turn activated a subset of the Hox genes, namely numbers six and up, which are the ones expressed in the trunk and tail. Harv and Betsy had done a number of experiments that involved making tadpoles without heads, or without tails, by manipulating the expression of these genes. Overexpression of an FGF, Cdx gene, or a posterior Hox gene would all cause the embryo to develop with no head, and so consist of an isolated trunk and tail. Conversely, the inhibition of FGF signalling or of Cdx function would produce an embryo that was just an isolated head, with no trunk or tail. These experiments illustrated very graphically the fact that we do know enough to suppress body parts in vertebrate embryos.

Thus far was science fact. The science fiction, or informed speculation, went like this: Suppose someone needed an organ transplant. You could culture some of his cells. They could be any type of cell, as the genes in all cells are the same; so white blood cells would do. Genes would be introduced into the cells while they were growing in culture that would have the effect of suppressing the development of most parts of the embryo, except for the part of the body that you want. This is almost achievable today as many of the organs required for transplantation (lungs, liver, pancreas) lie in the foregut region and would remain intact following suppression of the development of head and tail using genes that we already understand. A genetically modified cell would then be fused with an enucleated human egg, and the resulting reconstituted embryo grown up, preferably in vitro, as an "organ culture." Particularly if the culture could be nourished at a higher rate than a normal embryo, it would grow to transplantable size within months. The patient would then have an organ graft that was a perfect genetic match and required no immunosuppression.

This is a plausible vision of what may one day be done if it really is possible to clone humans from tissue culture cells. It may sound good if you happen to be an individual needing a transplant, but from the public relations point of view, it is a lethal cocktail of all the things that make people feel uneasy about modern biology. You have genetic engineering, manipulation of human embryonic material, animal experiments, and high-tech medicine all rolled into one. Although I would not be opposed in principle to pursuing work along these lines, it is an ethical minefield and would certainly need careful regulation. The least plausible part of the scenario is the growth of the organ culture in vitro. In vitro culture of mammalian embryos is painfully difficult, and we are a long way from being able to support any type of mammalian embryo through gestation outside the mother. On the other hand, fertilized eggs are being implanted in the wombs of women every day, so my guess is that when the time comes, female volunteers will be needed to incubate the organ cultures. This could be an act of love, to grow an organ for a relative who needs one. But it could also be an act of commerce, and the unseemly rows arising from existing maternal surrogacy cases make one doubtful about how wise it would be to go down that road.

Horizon's Visit

When the *Horizon* team came, we pulled out all the stops to give them a good time, the whole lab helping to put on nice-looking visual displays. The producer was Debbie Cadbury. She was very lively and enthusiastic, and on arrival she explained to me that the whole filming schedule had to be arranged around the meal and coffee breaks for the camera and sound crews. This sounded a bit like the old-style British working practices that had been so effectively suppressed during the Thatcher terror. I was then surprised to learn that the camera and sound men were not BBC employees, as the BBC had recently made all such technicians into self-employed contractors. But the old working practices had not yet changed much, as they were all still being hired fairly regularly. I wondered what things would be like in a few years time when many new young technicians had joined the pool and were competing for a limited number of programme slots. Debbie held a budget for making the programme and had to use this to pay the crew and also to pay for the film. This last point was of some strategic importance. Unlike news reporters, who always use videotape, the makers of quality documentaries like *Horizon* often still use real photographic film. This is very expensive, and when it has been used, it is finished. So once the producer has committed herself to a day's filming, the process is irreversible, and she is more or less committed to using the material. Of course, the editing is very heavy, so a day's filming will end up as just a few minutes of actual air time.

Although we had prepared several venues for filming, in which different aspects of our work could be demonstrated, the whole morning was spent arranging the lighting for the interview of myself. Documentary makers can be absolutely obsessive about the lighting, and no trouble is too great to get it right. So once it is right, it is easier to move all your personnel and equipment to the good light than to contemplate arranging the lighting again somewhere else. I had rehearsed the interview very carefully and discussed it with Jane Bufton, the University Public Relations manager. She is a vivacious Welsh lady who had previously had a much tougher job running public relations for the nuclear reprocessing plant at

Sellafield. She had reassured me that *Horizon* was a responsible programme and reminded me of the main rule that I had learned at the SDS:
when on television, never try to say more than three simple things. Within
this constraint, my agenda was to say as much as possible about our own
work and as little as possible about the sensitive area of human organ
transplantation. However, my preparations were somewhat undone by the
numerous repeat takes. Being new to television, I found it disconcerting
that the cameraman didn't necessarily tell you when filming was going on
and when it wasn't, and I also found it difficult to know what I had already
said, because in the numerous repeats of parts of the interview the questions were not always the same each time.

All journalists love moral anguish and controversy, and Debbie was no
exception. So she tried hard to get me to say how important it was to push
forward the project of growing human organs using the new cloning technology. But I would not be drawn. In fact, I backpedaled somewhat from
the time of the initial phone call and now stressed the highly sensitive nature of anything involving human embryos and the need for strict regulation. At one point, with the camera firmly off, I had said to her that I considered the moral problems raised by embryo research did not approach
the everyday moral dilemma involved in ordinary abortion. She tried to
get me to say this again in front of the camera, but I refused. I knew that
to even mention abortion in the context of this programme would be to associate myself indelibly with it, and it would then be difficult to disentan-
gle any sort of scientific message.

Once the interview was over, I had to endure being filmed at the microscope for about an hour. There were innumerable repeat takes of me turning
the knobs, moving the lenses, and moving the specimen. Then we put Betsy
in front of a dissecting microscope, demonstrating microinjection of embryos using a blue dye. Of course, genes are not really blue, but you need
something that you can see on television. By the end of the day, the film was
made. I was a little apprehensive about how the interview would turn out,
but I was happy with the visuals. I knew that only a couple of minutes would
eventually be used but felt sure that the pictures of Betsy would look good.
Also, because all the expensive film had been used, we felt that they would
find it hard to leave us out altogether. So we were content, and I could mentally tick off another small ambition that had been fulfilled.

The Explosion

I knew that the subject matter of the *Horizon* interview was sensitive,
because it involved so many areas that are themselves individually sensi-

tive: human embryo research, genetic engineering, animal experimentation, and transplantation. Just in case the Animal Liberation Front decided
to come and blow me up, I decided it would be wise to remove the name
plate that advertised my car parking space. But apart from this precaution,
I thought little about the matter until a week before the programme was
due to be shown. On that day Debbie rang me up. It was to warn me that
a *Sunday Times* journalist called Steve Connor was writing a preview to the
programme and would probably phone. "You can trust him; he is okay,"
she said. He did phone, and I said to him much the same things that I had
said on the *Horizon* interview. The next day a press photographer came
round and took hundreds of photos of me. So I knew that there would be
an article in the *Sunday Times,* which is the highest-circulation quality Sunday paper in Britain, and I knew that there would probably be a picture of
me. But I had the impression that this article was just some sort of preview
of the television programme, to appear somewhere in the depths of the vo-
luminous newspaper.

On Saturday night we were away from home at a dinner, and I went out
on Sunday morning to get the paper. I was pretty surprised to see myself
not in some obscure television supplement but on the front page! Worse
still was the headline: "Headless frog opens way for human organ factory."
The article was actually quite well written and did not significantly mis-
represent what I had said. But it did lay stress on the headless tadpoles we
had made and somehow managed to give the impression that the next step
was headless humans, something I had been very careful to avoid on the
Horizon interview. I was in for my baptism of fire at the hands of the media, but being at the time still ignorant in the ways of the world, I had no
idea what was in store. It took us a long time to get home, as there was a
jam on the motorway. Shortly after we got back, a friend of mine phoned
and said he was disturbed at the story and that I was in for a lot of trouble.
This rattled me a little, but I tried to brush it aside. We unpacked and had
our lunch, and it wasn't until then that my wife found a note pushed
through our cat door. It was from Jane Bufton. "Please phone immediately"
it said. "We have requests for interviews from Sky, HTV, CBS, ABC, etc.,
etc." Jane didn't know my phone number, but she had been able to find
out where I lived. In this regard she had done better than the HTV crew
who apparently had scoured the area looking for me and had even enquired at a pub where I had been on the Thursday evening! Fortunately,
they had not found me, as I would not have relished a posse of paparazzi
outside the house. Jane gave me some advice on the relative merits of the
various people who wanted to see me, based partly on the likely social
class of the audience and partly on whether the appearances were straight
news or might involve debate with hostile opponents. I made an appoint-

ment to do an interview the next morning on *Today*, the BBC radio national breakfast news show. Then I arranged with Sky TV to go and be interviewed that afternoon at the university. In some ways it was like a rerun of the *Horizon* filming, except that I had to arrange all the demonstrations myself and it was much quicker. These were newsmen, not documentary makers, and they were in and out within the hour. I spoke for a while on the phone to the *Daily Telegraph* and then hours more talking to lower-priority journalists who had managed to find my office number.

On Monday I arrived in Bath at 6:50 A.M. to do the *Today* interview. The BBC studio in Bath is not the most impressive media hub in the world. It consists of a two-room flat in a Georgian building, which most of the time is unstaffed. In order to get in, the BBC has to contact the "key holder," who is a self-employed person living nearby. There was nobody about when I arrived, and I started to worry that I would not be able to get in. Fortunately I had brought my wife's mobile phone, and I called *Today* in London, who assured me that the key holder would arrive within minutes. Indeed she did, and she sat me down in front of the microphone and provided me with a glass of water to deal with the frog in my throat, which, perhaps appropriately, was at that time giving me a lot of trouble. This interview was live. I had by then learned enough to realise that this was an advantage, because in a live programme, anything you say goes out on the air, unedited. So I decided to start with a plug for the university. It so happened that a few weeks previously a college in Bath had started calling itself "Bath Spa University College." This irritated the university, and a dictate had gone out to the effect that we were "The University of Bath," not "Bath University," as an attempt to avoid confusion between a high-status research university and a mere teaching college. I knew that the presenter would be unaware of this fine marketing distinction, and sure enough he led in with, "In our Bath studio is Professor Slack from Bath University." I had often heard politicians butt in at the beginning of their contribution with an "If I may just correct something" interjection that enabled them to plug something they wanted. "If I may just correct something," I said, "it's the University of Bath, the top research university outside Oxbridge and London, not Bath Spa University, which is a different institution." My claim was based on the 1996 Research Assessment Exercise, in which Bath did, in fact, rank just below Oxford, Cambridge, and London. Probably very few of the audience noticed, but the faculty at the university were very pleased with this bit of mischief. I knew that the next rule of media management was to deliver my own message regardless of the questions. Because it was live, this was easy. I talked for two or three minutes about our experiments and left the presenter only the briefest of opportunities to ask his own questions about organ transplantation. Unlike a recorded interview, which can go on forever, a live one has to be truncated by the 7:30 news headlines, or whatever, and so I escaped unscathed. Then Maggie, the key holder, asked whether I could fit in another interview with Radio Bristol, the local BBC radio station. I agreed and in another few minutes was on the air again. This was a bit tougher, as I was asked about headless humans, but I made it quite clear that I and other scientists were dead against producing any such thing. I left just as the dawn was breaking and still before the morning traffic had filled up Bath's elegant streets with fumes. I felt "so far, so good."

When I got to the university, it was a long chat on the phone to Sue Walters of the *Bath Chronicle*. Obviously, it is important to look after your own local media. A photographer came round and took several more photos of me sitting at microscopes or gawking at the frogs. Then the real onslaught started. My poor secretary, Heather, had the whole of the world's media bearing down on her. I made appointments to see HTV, the local television company, and ABC, a big U.S. network. While I was speaking to ABC on the phone, he said, "You're on Sky right now. Now I can see what you look like." Then I had to shoot back into the centre of Bath to do a rush interview for Radio 5, a rather down-market but national BBC radio station. Maggie opened up the studio again. She had obviously not seen so much activity for a while. Despite Bath's fame as a cultural centre and the base of one of the best rugby teams in the country, it doesn't produce much news. Apparently, the Bath studio often lies unused for a week or two at a time. The questioning on Radio 5 was a bit more aggressive along the lines of, "Isn't all this contrary to Nature? Where is it all going to end?" Again, I firmly squashed any suggestion of headless humans and felt that things had not gone too badly.

Then it was back to the university to face the television companies. By this stage things had become so chaotic that I insisted that Jane and her Public Relations department must handle the bookings themselves. Although my secretary had borne up well under the media onslaught, the tension in the office was rising to dangerous levels. The highest drama normally reached in a departmental office concerns such things as why the stationery cupboard is bare, or how on earth can I get my computer to behave. This was something quite new. The secretaries really believed that it was all very dangerous and that bomb-throwing militants would burst in at any moment, motivated by the deadly cocktail of human embryo research, animal rights, and genetic manipulation. I insisted that Public Relations should not only filter the calls for seriousness but also choose who should be allowed in and restrict the number of interviews to a manageable level. Once this principle was established, they started doing a good job. But Jane reminded me that we were in desperate need of a press release. Because we had not initiated any story, there was no press release! She had thrown one together herself based on the *Horizon* material, but a quick eyeball suggested to me that it was not suitable, as there was too much about organ transplantation. So in between television interviews, I wrote one myself and had it ready by about 1 P.M. It was apparently very useful, as most of the callers to Public Relations were satisfied when they received it and did not bother us again. Jane herself was very concerned about the "headless frog." Of course, there were no headless frogs, only headless tadpoles, but this distinction had not been apparent to the subeditor who wrote the headline in the *Sunday Times:* "Headless frog opens way for human organ factory." Jane felt that nobody would care about headless tadpoles, but headless frogs could provoke an animal rights backlash. I was more concerned about the headless humans. They had no more reality than the headless frogs, but I had realised that this was a potent image that the media were going to propagate. Logically speaking we should not be worried about the possibility of headless humans. After all, no head means no consciousness, no sensation, no pain, no individuality, no humanity. But appearances are very important, and I knew that it would be suicidal to advocate the creation of anything that looked remotely human for a purpose such as transplantation. On the recorded interviews, I did my best to prevent any questions about headless humans; on the live ones, I just had to say that I was strongly opposed to the possibility.

That afternoon we ended up doing no fewer than four television interviews, each with associated footage of frogs and tadpoles, people sitting at microscopes, and people pretending to pipette samples at the bench. I had taken the precaution of wearing a jacket and tie, as I know that if you are saying something outrageous you should dress conservatively. On television much of the message lies in your appearance and body language; the audience would know that a solid and reliable person would never create headless humans. The opposite holds equally strongly; if you are a guru doling out some boring platitude, you should dress in a more casual and adventurous fashion. First, there was HTV; then a local BBC team arrived without an appointment. I almost threw them out but felt that we should really try to accommodate the BBC, so I let them stay for a bit. Then ABC arrived, and finally CBS. The American reporters were very professional and also appeared remarkably interested in our work and specimens. This is somewhat surprising when you consider the range of stories they covered. The previous assignment for the ABC team had been a bullfight in Spain!

As the first rule of media handling is to say no more than three things and say them repeatedly, all my interviews were exactly the same. The first question is "Explain very simply what you have actually done." I explained that we were working on how the head-to-tail pattern of the embryo developed. We had identified some genes whose function was to form the trunk-tail part of the body. If you put too much of these genes or gene products into a frog egg, then the tadpole would form just a trunk and tail, with no head. If you inhibited the action of these genes, then the egg developed to form an isolated head. The next question is always "Would it work in humans?" The answer is "Almost certainly it would, because we know from many people's work that the basic developmental mechanisms in all vertebrates, fish, amphibia, birds, and mammals are very similar indeed." Then things started to get tricky as they asked about organ transplantation. I explained that developmental biology is a very basic science and will have numerous applications in wide areas of medicine: cancer, diabetes, arthritis, neurodegeneration, and I tried to fit in my quotation, "You can't fix it if you don't know how it was put together in the first place." But journalists aren't paid their large salaries to be fobbed off with this sort of generality, so they always wanted to know more about organ transplantation. I explained that this was one of the possible applications of developmental biology, combined with the newly discovered ability to clone mammals. It was far in the future and would need very careful consideration of the ethical and legal issues. I explained the possible procedure whereby organ cultures might, one day, be created. When asked about regulation, I said that serious ethical problems were raised here and that an agreed regulatory framework would be required. I argued that such measures had been successful when dealing with possible hazards of recombinant DNA technology, animal experimentation, and the existing regulations on human embryo research in the United Kingdom. "How far would you go?" they asked. I had had time to think about this since the original discussions with *Horizon,* so I had made up my own mind. "I should be opposed to creating anything that looked like a human embryo, or any procedure involving maternal surrogacy." I have wondered since whether these scruples are sensible given that people are already being murdered for their kidneys and that things are going to get worse and not better in the coming years. However, the story was already quite strong stuff for the general public, so it was probably better to be cautious and conservative.

Each television team was difficult to shift as once they had set up their lights and cameras, they insisted on "Just one more shot of . . . someone pipetting; Slack talking to someone; the tadpoles." It is all acting of course, but the media make it reality. The U.S. teams seemed to have cameras whose scan rate made them unable to film pictures on our monitors. This was a handicap, because all the action in developmental biology occurs down the microscope, and the easiest way to display this to an audience is to put it on a monitor screen. So we printed out some stills, and I had to let them film each still and give them a verbal caption. We finally got rid of the last team about 6:30 P.M., and I staggered off for a drink with a couple of people from the lab.

The next few weeks showed persistent media interest but at a much lower level. After 24 hours a story is old news, and the big networks move on to something else; so it is just the specialist science programmes that remain interested. I had a bad moment on Tuesday when it seemed that one of the things the big networks were moving on to was a story from Hong Kong about organs from prisoners executed in China being sold on the black market. This seemed a bit close, but it never caught on in the U.K. media, so I don't know whether it ever got linked to our story.

There were several more television and radio interviews over the next few weeks. The most difficult were the Canadian ones. The Canadian companies are evidently not quite as rich as their American counterparts, and they do not maintain a full corps of foreign correspondents. So they just send a camera crew and do the interview down the phone from Canada. By then I was a sufficient enough veteran of television interviews not to be worried that the interviewer was 6,000 miles away. But the phone connections did cause trouble. Our labs are all lavishly wired up with ethernet cabling for fast computer networking, but they do not have phone points! The first time, I had prepared to do the interview in the microscope room, so we had to run a phone line across from the nearest office and hope that no one else called my colleague at the critical moment. This caused so much trouble that for the next one I insisted that the interview be done in my office, thinking that this would solve the phone problem. But television crews always want to film in the lab, a more exotic backdrop for most viewers than a boring old office. I had to mollify them by producing my collection of pickled axolotls in glass jars, which were tastefully arranged to one side of me to provide a little colour and interest.

That week I also had a discussion with a reporter from *Science*. This assignment differed from all the others because I was feeling rather apprehensive about what my scientific colleagues would think of it all. Among professional scientists, the cardinal sin is to release a news story to the media about a piece of work before it has been published in a proper scientific journal. Fortunately, I had not done this. Two papers had already been published on the FGF-Cdx-Hox pathway, and although a third was just being written, the interviews I had given were of such a general character I could legitimately claim just to be describing the contents of the first two. So far so good. But the already-published status of the work brought another problem. Anyone in the field would be very puzzled because they would know that there was nothing new about the headless tadpoles. Headless tadpoles are a not uncommon result of many experiments with

Xenopus embryos and had been made many times before. So it was important to establish the fact that I had not released any story myself and claimed some sort of breakthrough when there was no breakthrough. Indeed, I had not launched the story myself. The media had come to me and asked me about it. If the word "breakthrough" was ever used, it was not by me. You could argue that I should have told all the reporters that there was really no story. I should have said: "Headless tadpoles are not new, and they have nothing to do with organ transplantation." But I could not quite bring myself to do this. In fact, I did get a few enquiries from science journalists who knew that headless tadpoles were not new, and I confirmed their belief. But I saw no particular need to volunteer this information to people who were arriving to talk to me about our work, which was, after all, a quite interesting and a respectable piece of science.

But when the reporter from *Science* rang, I had to tell him the truth, because *Science* would be read by my colleagues. I told him that we had released no story, that the story was not new, that the link with organ transplantation was long-term speculation. I explained that it was all an accident arising from the positioning of the *Horizon* preview article on the front page of the *Sunday Times*. At this time I also wrote a letter to *Nature*. Their editorial on that Thursday contained a brief reference to me and, horror on horror, an indication that I was at the University of Bristol! Was Bath's hard-won international fame to be lost so soon? I used this error as a pretext to write my letter, with the same explanation as supplied verbally to *Science*. I hoped that these two initiatives would limit the damage done to my professional reputation, which might be severe if it prejudiced people against my grant applications. Fortunately, *Nature* published my letter verbatim, and *Science* ran a half-page feature about media frenzy over the headless tadpole. This was good because it made it clear that the story had originated from the *Horizon* film and was not the result of a news release from Bath. It also referred to me as "widely respected," and included my modest quotation, "There's absolutely nothing special about our work compared to work in many other laboratories." I felt this would serve as useful protection against other scientists irritated that such a fuss was being made about a good but unremarkable piece of work.

I then had an anguished e-mail enquiry from Matt Scott, chairman of the American Society of Developmental Biology (SDB), and incidentally one of the discoverers of the homeobox (Chapter 8). "What the hell's going on?" he said. "I'm not giving any press conferences until I check the facts with you." I sent him an e-mail with essentially the same story as supplied to *Science* and *Nature*. By return I found that the SDB was calling for a five-year moratorium on human cloning. They were very worried that the media coverage would cause a panicky U.S. Government to bring in

some repressive legislation that banned large swathes of biological research. The previous experience of recombinant DNA technology suggested that this might be forestalled by a voluntary moratorium by scientists themselves. They wanted the British Society for Developmental
Biology to do the same. At the time I happened to be the national secretary
of the British society myself. So it was easy for me to agree to put this item
of business on the agenda for our next committee meeting. I was unen-
thusiastic about the initiative, however, because knew that our society only
had about a thousand members and included rather few of the human embryologists who might have the opportunity to do this sort of work with
human embryos. I also knew that the American society is even less important in the context of U.S. science than we were in Britain, so a statement
from these two societies would carry rather little weight and might start
another media explosion to no useful effect.

While on the matter of e-mail, it is interesting that the world's media
made relatively little use of the internet during this business. The University of Bath has an internet site, and I have a page. Anyone could find me,
my picture, my e-mail address, my phone number, and something about
my work, including a picture of a headless tadpole, from this source. In
fact, relatively few did. I answered perhaps 20 e-mail enquiries from journalists, while Public Relations dealt with hundreds of phone enquiries. I
conclude that, despite all the hype, the age of the internet is only just
dawning and that our communications will be unrecognisable in 10 years
time.

Meanwhile, the story had run round the world. Nowhere was too remote to feature the headless frog. It was in local newspapers in small corners of the American Midwest; it was in papers in Seoul, Korea; it was in
papers in Brazil; it was in Russia; it was in the Hebrew press in Israel (in
this paper I had the distinction of appearing next to a story about a wed-
ding in Taiwan conducted in the nude!). Anywhere around the world
where I had a contact, I could be sure that they would have seen the story.
The apotheosis came on the Saturday night, on the British satirical television programme *Have I Got News for You,* always a feast of highbrow bad
taste. Sure enough, the headless frog made a brief appearance, and after
that the general consensus was that I had well and truly arrived among the
literati.

The case of the headless frog had taught me several new things about
communicating our sort of science to the public. First, the idea that the developmental mechanisms of all animals are much the same, as discussed in
Chapter 8, is now familiar to all developmental biologists. But it is not familiar to anyone else. It is extremely important because it means that results obtained on zebra fish, frogs, or mice are almost certainly also true for

the human embryo, so it should always be included among the three "bullet points" that a developmental biologist should make when in front of the cameras. Second, we take for granted that it is okay to perform experiments that result in the production of funny-looking embryos, because we know people have been doing such experiments since the time of Spemann. But the general public does not know anything about classical experimental embryology. To them, if a headless tadpole is in the news, then it is a novelty, and some people may feel that such experiments are cruel to tadpoles and should be stopped. It is a sobering thought that when Spemann received his Nobel Prize, the publicity would have been restricted to a few serious articles in a few serious newspapers, but when a science story hits the wires today, it can be flashed to billions of people in a few minutes. Finally, in these sensitive areas, appearances are very important. As explained above, I find it difficult to argue rationally against the creation of a headless, but otherwise intact, human embryo, for transplantation purposes. But I have a strong emotional feeling that it would be wrong because it would represent a violation of the human form, and I know that many other people will share this perception. The feeling that external form should be respected is likely also to extend to animals. My guess is that genetically engineered farm animals produced by cloning technology will be commonplace in a few years time. But they will still look like perfectly normal cows, sheep, and pigs. A distaste for interference with external appearance may be one of the strongest limitations on the applications of developmental biology in the next century.

So where would it all lead? Would I be ostracised by the scientific community or would I become a media celebrity and not have to worry about research funding again? Time will tell. But from a strictly local point of view, the event had undoubtedly been a success. Had the University of Bath purchased advertising time on the world media to equal the exposure, I am sure it would have cost several times my lifetime salary. Although nobody had planned it this way, the portion of the global audience who were not professional developmental biologists (around 99.9999 percent) would presumably have concluded that some important discovery had been made at the University of Bath. If this is the sort of university where important discoveries are made, well, it must be a good place!

Epilogue

So what is science? Philosophers may be concerned about the nature of scientific method, or whether falsification is better than verification, or why there seems to be a correspondence between mathematics and physical reality. Politicians see science as an engine of economic growth and are always trying to make academic scientists do things that are more "relevant" and closer to the market. Students see science as a large body of received wisdom that must be learned and mastered in order to get a job. The public has some respect for science but also fears it, being, for example, concerned about all those thousands of deaths that it thinks have been caused by radioactivity and genetic engineering.

Working scientists rarely worry about such things. Their concerns are much more focused. They want to get their paper into *Cell*. They want to get their own lab. They want to get grant money. They want to get tenure. And when they have all these things, they want to be a member of the National Academy of Sciences or a Fellow of the Royal Society. To get what they want, they will scheme, intrigue, manoeuvre, and bargain like anyone else. In other words, what scientists really do is take part in a competitive rat race similar to that in most other branches of human activity.

The remarkable thing is that in the course of doing all this, they also manage to transform the world. My own view is that the special character of science lies not in a universal "scientific method," which it is presumed to practice, but rather in its sociological characteristics. Consider whether there is any method common to astronomy, particle physics, meteorology, paleontology, and molecular biology. Some of these subjects are mathematical, others are not. Some are experimental, others are not. Some have repeatable objects, and for others the object is a unique historical sequence in which the chain of cause and effect can never be established with certainty. On the other hand, they all, without doubt, lie within the natural sciences. We may hesitate a little over admitting the same of economics or archaeology or history. But the practitioners of these disciplines are also attempting to advance knowledge and to answer their own particular empirical questions in their own way.

Perhaps what is really characteristic of the natural sciences is that they focus on very small questions that are capable of being answered. Every so often the answers to a number of little questions become a single answer to a big one, or maybe even cause a new big question to be asked for the first time. For example, the study of a tumour extract that made nerve cells grow in chick embryos eventually led to the discovery of all the modern families of growth factors and hence to the creation of a significant sector of the biotech industry. The pursuit of the genetics of the four-winged fly eventually led to the understanding of the molecular basis of the head-to-tail pattern of all animals and explained to a large extent how all animals develop.

But does the nature of science really matter? I think it does. It is true that developmental biology has had relatively little impact on society in the twentieth century, except for a few innovations such as in vitro fertilization. But there is little doubt that it will loom large in the twenty-first, and that the practical applications of developmental biology will create many moral dilemmas. We shall without doubt eventually have the power to introduce directed genetic changes into the human germ line or to grow human kidneys in bottles. These technologies will have great power for good or ill and will need careful handling. The more understanding society has of the basic science, and the people who do it, the easier it will be to get the regulatory framework right.

For the young scientist starting his or her career, the trick is to identify which particular little question, of the many millions currently capable of solution, is actually going to open up something interesting. There will always be an element of luck as well as judgement in this. Like racing tipsters, elder statesmen in science who attempt to give advice to their younger colleagues might be asked why they did not make their own fortunes by betting their careers in the directions they are now recommending. I am not really important enough to give "Advice to a Young Scientist," but I will do so anyway. My recommendations are very simple. Work on something that is presently an unfashionable backwater. The fashionable stuff always involves lethal competition and is sure to be mined out within a few years. Because you can never be sure which backwater will explode next, at least choose something that you yourself are really interested in. Then if you don't get your paper into *Cell,* or don't get tenure, or don't get lots of medals, you will at least have spent some time doing something that you knew at the time was important and felt was really worthwhile.

Bibliography

Some original scientific papers and a few secondary sources describing the work covered in this book are listed here. This short list is hopelessly subjective and biased, as the areas of science I have mentioned in the book are very wide and the number of important primary papers runs to many thousands. But it may serve as an entry point to the literature for those who are interested.

The experiment described in Chapter 1 is published as:

Slack, J. M. W., B. G. Darlington, J. K. Heath, and S. F. Godsave. 1987. Mesoderm Induction in Early Xenopus Embryos by Heparin-Binding Growth Factors. *Nature* 326:197–200.

The other three papers resulting from the 1987 gold rush for mesoderm inducing factors:

Kimelman, D., and M. Kirschner. 1987. Synergistic Induction of Mesoderm by FGF and TGFβ and the Identification of an mRNA Coding for FGF in the Early Xenopus Embryo. *Cell* 51:869–77.

Smith, J. C. 1987. A Mesoderm Inducing Factor Produced by a Xenopus Cell Line. *Development* 99:3–14.

Weeks, D. L., and D. A. Melton. 1987. A Maternal Messenger RNA Localized to the Vegetal Hemisphere in Xenopus Eggs Codes for a Growth Factor Related to TGFβ. *Cell* 51:861–67.

A useful guide to growth factors and their properties will be found in:

Heath, J. K. 1994. Growth Factors. Oxford: IRL Press.

Some recent guides to the biotech industry:

Kornberg, A. 1995. The Golden Helix: Inside Biotech Ventures. Sausalito, Calif.: University Science Books.

Maulik, S., and S. D. Patel. 1997. Molecular Biotechnology: Therapeutic Applications and Strategies. New York: John Wiley.

The organizer paper:

Spemann, H., and H. Mangold. 1924. Über Induktion von Embryonenanlagen durch Implantation von artfremder Organisatoren. *Arch. Microsk. Anat. EntwMech.* 100:599–638.

Spemann's book remains a fascinating read for those interested in experimental embryology:

Spemann, H. 1938. Embryonic Development and Induction. Reprint, New York: Haffner, 1967; New York: Garland, 1988.

The paper that started the first gold rush for the "organizer substance":

Bautzmann, H., J. Holtfreter, H. Spemann, and O. Mangold. 1933. Versuche der Analyse der Induktionsmittel in der Embryonalentwicklung. *Naturwissenschaften* 20:971–74.

A review of induction in the *Xenopus* embryo, written after the dust of the second gold rush had settled. It includes primary references to work on Wnt, BMPs, noggin, and chordin:

Slack, J. M. W. 1994. Inducing Factors in Xenopus Early Embryos. *Current Biology* 4:116–26.

The founding text on homeosis:

Bateson, W. 1894. Materials for the Study of Variation. London: Macmillan.

A good general guide to *Drosophila:*

Lawrence, P. A. 1992. The Making of a Fly. Oxford: Blackwell.

The genetics of the Bithorax complex:

Lewis, E. B. 1978. A Gene Complex Controlling Segmentation in Drosophila. *Nature* 276:565–70.

A brief account of the mutagenesis screen that led to the identification of most of the developmental genes in *Drosophila:*

Nüsslein-Volhard, C., and E. Wieschaus. 1980. Mutations Affecting Segment Number and Polarity in Drosophila. *Nature* 287:795–801.

A useful review on the Hox genes:

McGinnis, W., and R. Krumlauf. 1992. Homeobox Genes and Axial Patterning. *Cell* 68: 283–302.

The "zootype" paper:

Slack, J. M. W., P. W. H. Holland, and C. F. Graham. 1993. The Zootype and the Phylotypic Stage. *Nature* 361:490–92.

Evidence for the great inversion:

de Robertis, E. M., and Y. Sasai. 1996. A Common Plan for Dorsoventral Patterning in Bilateria. *Nature* 380:37–40.

Index